Cipriano Lialunga

Study of Slope Instability Situations in the Catumbela Area

Cipriano Lialunga

Study of Slope Instability Situations in the Catumbela Area

Slope Instability Situations

ScienciaScripts

SUMMARY

CHAPTER 1	**2**
CHAPTER 2	**4**
CHAPTER 3	**14**
CHAPTER 4	**31**
CHAPTER 5	**38**
CHAPTER 6	**58**

CHAPTER 1

INTRODUCTION

The study of slope instability processes in Angola, in general and in the municipality of Catumbela in particular, is an essential task due to the disastrous consequences these processes have for people and society in general, given the general increase in urbanisation and the development of areas subject to instability, as well as the deforestation of the respective areas for occupation and the possible increase in rainfall rates caused by climate change.

The occurrence of instabilisation is related to various causes, including increased urbanisation and the development of areas subject to instability movements; the continuous removal of vegetation from these areas and the occurrence of high rainfall, often related to climate change.

When landslides occur, they can have various consequences, ranging from economic and social damage to the loss of human life.

Natural and human-induced factors contribute to the occurrence of instability movements (Martins, 2007). Human-induced factors include actions to alter the geometry of slopes, such as the increase in space for building new homes and the construction of motorways, without observing the appropriate technical aspects or sufficient geological/geotechnical knowledge.

During the process of political and military instability, which lasted from 1975 until 2002, people who usually lived in the interior of the country moved to the coast and, as a result, the number of inhabitants in towns located on or near the coast increased significantly. The development of cities without proper land-use planning has led to the emergence of housing in areas adjacent to unstable slopes. The area of the municipality of Catumbela is made up of plains (near the seafront), plateaus and more mountainous areas, the latter with a maximum elevation of around 931 metres.

In recent years, communication routes have been opened up and houses have been built illegally and without proper planning, often in the vicinity or even on slopes susceptible to unstable movements. Most of the houses are occupied by people from the interior of the country. The rainy season, from January to April, increases the possibility of erosive processes and instability movements in the urban area of Catumbela.

This study selected slopes/verges in the Catumbela region for the analysis and study of instability situations. The sites chosen are in the neighbourhoods of Cambambi, Poli, Caputu, Tata, Chiúle, São Pedro and Luongo, which are located in the urban and suburban areas of the municipality of Catumbela.

In the chosen locations, there are instability events that can damage homes and communication routes, and possibly cause casualties, and Catumbela's municipal authorities should intervene in these areas. This study required detailed knowledge of various types of slope instability, such as landslides, falls, *toppling* and *flows*.

1.1 - Objectives

1.1.1 - General objectives

1. To analyse information on the occurrence of slope instability situations in the Catumbela municipal area.
2. Recognising situations of slope instability and defining their influence on the correct planning of population settlements and communication routes in urban and suburban areas.
3. To contribute to a better understanding of the mechanisms that condition instability movements, with a view to improving the management of public authorities in drawing up measures to prevent and mitigate the consequences of instability situations.

1.1.2 - Specific objectives

1. Identification of slopes in the Catumbela area;
2. Understand and analyse the situations of slope instability in the Catumbela municipal area;
3. Define the determining factors in the occurrence of slope instability situations.
4. Evaluate the consequences of slope instability situations in the municipality of Catumbela;
5. Establish measures to mitigate the impacts of slope or hillside instability.

In order to achieve the objectives set out in this work, it was necessary to collect, process and analyse data based on the following:

Research question/problem:

What factors influence slope instability in the Catumbela area, what are their types and consequences?

Population: Taludes/Vertentes of the Municipality of Catumbela.

1.2 - Structure and synthesis of the work

The work is structured in six chapters: the first contains an introduction and describes the general and specific objectives; the second describes the types of instability, taking into account the collection and analysis of the literature; the third characterises the study area, covering aspects related to the geography, climatology, geology and geomorphology of the municipality of Catumbela; the fourth describes the methodology adopted to carry out this work; the fifth characterises the different aspects of instability situations, carrying out a comparative study and analysing the results of the classifications used; the sixth presents the conclusions of the work, as well as recommendations for the correct development of the municipality of Catumbela. Bibliographical references are also presented.

CHAPTER 2

Description of the types of instability

The occurrence of slope instability often leads to accidents whose economic and social damage can be very high, particularly in urbanised areas. Changes in topography, hydrological and hydrogeological characteristics, as well as geomorphological processes, caused by the urbanisation of new areas, can lead to situations of instability (Coelho, 1979).

Instability movements are defined as the downward displacement of materials along a slope (Teixeira, 2005); Cruden (1991) defined instability movement as the downward displacement of a mass of rock, soil or debris on a slope.

The generic term landslides covers a range of different types of material movement associated with the action of gravity on slopes or hillsides. Instability movements occur naturally on the earth's surface as a form of relief modelling, which is the result of precipitation and erosion processes (Silva, 2011). Instability situations also occur due to anthropogenic action, which results in the modification or alteration of the natural characteristics of the terrain. Human action is evident in the opening of new roads or the building of homes, which can lead to changes in the original geometry of the slopes.

Landslides can be observed both on natural slopes and on excavation and embankment slopes. They are present in various engineering works, as well as in deposits of materials related to mining operations (Silva, 2011).

The study of landslides is the subject of study in various disciplines such as geology, geological engineering and civil engineering, so there can be differences in the definitions proposed or used by different professionals or researchers (Highland & Bobrowsky, 2008).

For slopes, there are different causes of instability, such as the action and presence of water. Another important factor is related to the height and slope of the slopes, as well as the orientation of the geological structures, in which the slope of the strata and discontinuities stands out (Bastos, 1999).

According to Small & Clark (1982), instability movements have an impact on the land affected and can, in extreme situations, endanger human lives, buildings and communication routes. Human activity on slopes is one of the main factors conditioning the processes, morphology and evolution of slopes (Guerra & Jorge, 2012).

2.1 - Classification of types of slope and slope instability

The classification of slope movements is of fundamental importance in applied geomorphology and land use planning, since defining the mechanisms at work is essential in the processes of controlling or correcting instability situations (Hansen, 1984).

Slopes can be considered natural, excavation or embankment. Slope movements take different forms, occurring due to the action of various factors and are related to causes whose identification can be difficult to establish.

There are various classifications of types of instability (Table 2.1). An initial classification for slope instability situations was defined in 1894 by Penck (Andrade, 2008).

The classifications of instability situations take into account different parameters such as the speed, the constituent material, the ruptures present, the causes, the morphological aspects and the duration of the movements.

In 1925, Terzaghi classified mass movements based on the movement of soils (dry state) and detrital materials, the latter associated with plastic behaviour (Andrade, 2008).

Sharpe in 1938 established a classification based on factors such as the nature and speed of the movements, the type of material, the water content and the causes of the instabilities. Coates (1977) focussed on the types of movement and the materials displaced. This classification agrees with Sharpe's with regard to the speed and size of the materials present, such as rock, regolith and sediment. As for the main types of movement, landslides (rotational and planar), flows and falls were distinguished.

Table 2.1 - Classification of movements on slopes/uplands adapted from Andrade (2008).

Author	Year	Criteria
Penck	1894	Types of mechanisms
Howe	1909	Types of material
Stini	1910	Types of mechanisms
Terzaghi	1925	Types of material
Heim	1932	Types of material
Sharpe	1938	Types of mechanism and material
Popov	1946	Age of the materials involved
Varnes	1958, 1978, 1984	Types of material, mechanisms and morphology
Hutchinson	1968, 1988	Types of material, mechanisms, morphology and speed
Zarub & Mencl	1969, 1976	Types of geological material and mechanisms
Nemock et al.	1972	Types of mechanisms and speed
Panet	1976	Types of materials and dimensions
Coates	1977	Material size and type of mechanisms
Sassa	1985	Types of material and mechanisms
Cruden & Varnes	1996	Types of mechanisms, material, speed and water content
Dikau et al.	1996	Types of mechanism and material

Zaruba & Mencl (1976) considered classes according to the geological material and the mechanisms present, defining movements of superficial deposits, rocky materials of pelitic composition, altered rocks and types of movements that are related to regional geological aspects. Sassa (1985) established a geotechnical classification related to the size of the material and the form of rupture.

Varnes (1958, 1978, 1984) developed a classification based on two main aspects: the type of material and the type of movement. The type of material was considered to be: rocks and soils (debris and earth). As for the type of movement or mechanism involved in the instability, six distinct types were defined: Falls, Toppling, Slides, Spreads, Flows and Complex Movements.

The most widely recognised classification of instability movements worldwide is that proposed by

Varnes (1978). Also noteworthy are the classifications of Varnes & Cruden (1996), in which characteristics such as slope activity, speed and water content were added, as well as that of Dikau et al. (1996), in which the classification of Varnes (1978) and WP/WLI (1993) were taken into account.

The speed of movement was also taken into account in the Cruden and Varnes (1996) classification (Table 2.2).

Table 2.2 - Classification of the speed of instability movements according to Cruden & Varnes (1996).

Speed description	Speed values	Type of movement related to speed
Extremely fast	>5 m/s	Collapse
Very fast	>3 m/min	
Fast	>1.8 m/hour	Sliding
Moderate	>13 m/month	
Slow	>1.6 m/year	Creep
Very slow	>1.6 mm/year	
Extremely slow	<1.6 mm/year	

Hutchinson's classification (1968) uses the term "mass movements", which includes slope movements (Teixeira, 2012).

In Hutchinson's first classification, the mode of deformation is not used as a parameter, but the original classification includes terms such as creep, landslides and movements related to the freezing and thawing of the surface (*frozen groundphenomena*) (Mattos, 2009).

Hutchinson's second classification, published in 1988, consists of seven types of movement: setbacks, *creep,* mountain slope ruptures, slides, debris movements in the form of flows, toppling, falls and complex movements (Mattos, 2009).

2.2 - Types of slope/verge instability movements

This study used the classification and designation of slope movements according to the proposals of Varnes (1978) and also considered by other authors such as WP - WLI (1996) - the UNESCO working group, Cruden & Varnes (1996) and Dikau et al. (1996), which can be seen in Table 2.3.

Table 2.3 - Classification of slope/ridge instability movements, adapted from Dikau etal. (1996).

Types of Movement	Type of material		
	Rocky substrate	Soils	
		Essentially coarse	Essentially thin
Collapses	Rockfall	Debris collapse	Soil collapse
Basements	Rock tipping	Debris tipping	Ground tilting
Rotational slides	Rotational sliding of rock material	Rotational sliding of debris	Rotational sliding of soils
Translational landslides	Translational rock slide	Translational debris slide	Translational landslides and mudslides
Lateral expansion	Lateral expansion of rock material	Lateral expansion of debris	Lateral expansion of soils
Fluency	Fluidity of rock material	Debris flow	
Complex	Combination of movements in rock material	Combination of movements in detrital material	Combination of movements in soil-type material.

2.2.1 - landslides

Landslides, which include falling blocks of rock, correspond to rapid to extremely rapid movements of rock and/or soil material, in which the unstable materials move downhill along very steep slopes (which can be vertical), and in which part of the movement takes place in free fall.

Landslides may correspond to the separation of blocks in rock formations, which is related to the presence of fractures. The possible movement of these blocks is conditioned by the frictional forces along the fracture plane(s). This implies that for movement to occur, the forces promoting instability must overcome the frictional forces, causing the blocks to move and consequently fall freely (Braathen et al., 2004).

All or part of landslides comprise a succession of events: rupture and detachment - free fall - rebound - rolling - immobilisation (Lamas, 1998). Unstable materials, such as rock blocks, travel a distance that is conditioned by various factors such as the slope's inclination, height, the shape and size of the blocks, as well as the amount of energy absorbed by the vegetation cover during the impact with the ground (Lamas, 1998).

Rapp (1961) and Hutchinson (1988) differentiated between primary and secondary falls, the former corresponding to the fall of rock blocks that detach from the parent rock. Secondary falls involve materials that have previously undergone instability movements, and comprise materials that have been temporarily trapped following a previous movement from a higher elevation. The main causes of landslides include: erosive processes whose action makes the most resistant formations stand out, the action of decompression forces, the action of roots that can dislodge the material that makes up the slopes, thermal variation throughout the day, percolation and water pressures, as well as the effects of freeze-thaw (Andrade, 2012). Collapses can also be caused by the action of watercourses and undulation that erode the base of slopes or cliffs (Bloom, 1970) (Figs. 2.1 and 2.2). Anthropogenic action is also noteworthy, particularly in engineering works such as the construction of roads or buildings.

Figura 2.1   - Landslides on the west coast of Portugal (Andrade, 2012).

Figura 2.2   - Fall of blocks in Pipas Bay (Namibe) (Andrade, 2009).

2.2.2 - Basculings

In tilting, there is a rotational movement of rock, debris or soil around an axis located below the centre of gravity of the unstable material. The type of movement is related to the presence of discontinuities with a high gradient and in the opposite direction to the slope face (Fig. 2.3). The movement has very variable speed values, from extremely slow to extremely fast. The volumes of material involved in these types of instability can be very significant, sometimes totalling around 1 km^3 .

According to Goodman and Bray (1976) the tilting of rock material can be classified as flexural, blocky, mixed and secondary, the latter types being associated with excavations, erosion processes and mining (Andrade, 2008).

Some of the main causes of tilting are related to load actions, decompression of the massif, water percolation and the presence of vegetation. This movement can occur due to the eventual removal of base material from slopes or cut slopes, which can lead to tipping situations.

Figura 2.3   - Tipping, taken from AHI (2014).

2.2.3 - Landslides

The term landslide, in general terms, can encompass all movements of instability that occur naturally or caused by man, with the exception of situations of ground subsidence (Favaretti, 2011). The WL/WLI working group (1993) defined landslides or slides as downward movements of a mass of rock, debris or soil that occur along rupture surfaces or zones of intense tangential deformation with a reduced thickness.

Landslides correspond to movements in which most of the unstable material moves as a coherent or almost coherent mass, with little internal deformation (Highland & Bobrowsky, 2008). Landslides can be of the rotational type (Figs. 2.4 and 2.5), in which case the rupture surface is curved and has an upward concavity, or of the translational type (Figs. 2.6 and 2.7), in which case movement takes place along one or more flat or undulating surfaces, sometimes with small rotational movements (Ayala-Carcedo & Posse, 2006). Translational landslides in rock masses include planar and wedge landslides.

Zêzere (2000) considered that in a landslide the mass displaced in the course of the movement remains in contact with the unaffected underlying material, showing very variable degrees of deformation according to the type of landslide.

The causes of rotational landslides are erosion caused by precipitation, marine abrasion at the base of cliffs, seismic action, increased water pressure, excavations on the slopes carried out by human action and increased load on the top or face of slopes.

Circulating (Rotational) Slides

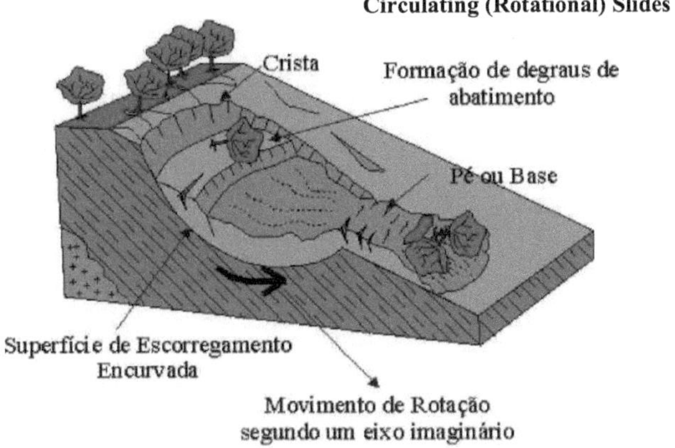

Figura 2.4   - Representation of a rotational slip adapted from Infanti Jr. & Filho (1998)

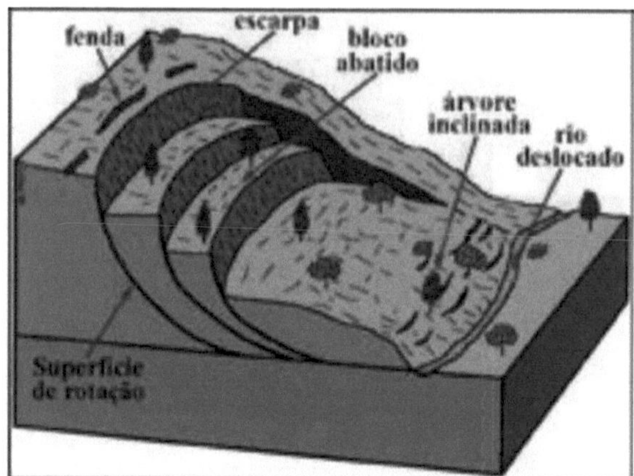

Figure 2.5 - Representation of a rotational slip adapted from Dias (2006).

Translational landslides correspond to moderate to fast movements of rock material, debris or even soils that are related to discontinuity surfaces such as stratification, schistosity, diaclases, fault planes and contact surfaces between different degrees of alteration.

Planar Slip (Transition)

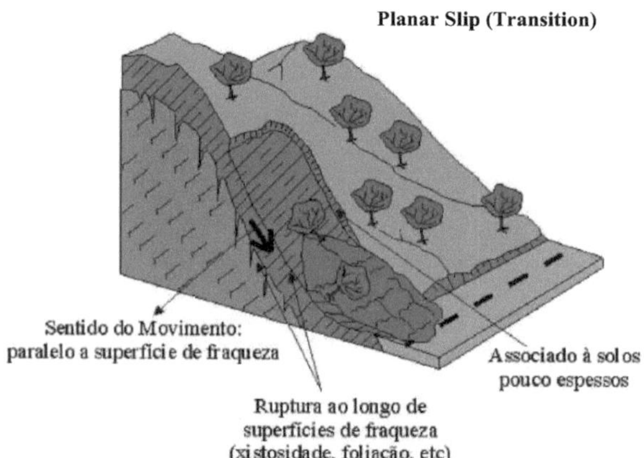

Figura 2.6 - Representation of a translational slide adapted from Infanti Jr. & Filho (1998).

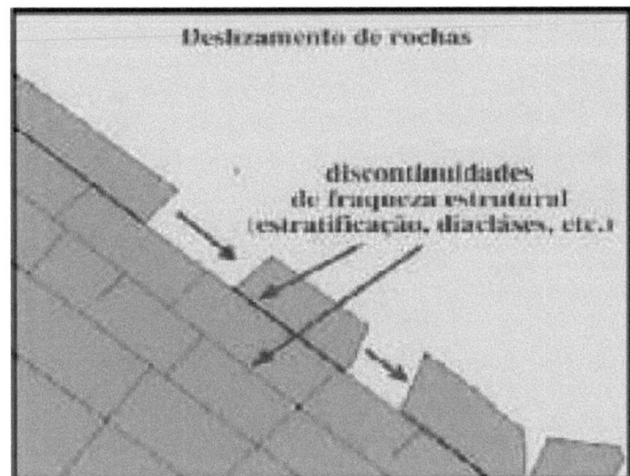

Figura 2.7 - Representation of a translational slip adapted from Dias (2006)

Translational landslides often occur due to human action on the slopes during the construction of roads or other infrastructure. Generally speaking, the main factors that influence translational landslides are lithology, the orientation of the slopes and embankments, the characteristics of discontinuities, seismic activity and the action of water, the latter being very important as it increases the specific weight of the soil and/or rock materials and causes a reduction in shear strength (Rodrigues & Zêzere, 2003).

2.2.4 - Lateral expansion

Lateral expansion can be defined as the "extension of cohesive masses of soil or rock, combined with a general subsidence of the fractured mass of cohesive material into softer underlying material. (...) It can result from the liquefaction or draining of the underlying soft material" (WP/WLI, 1993).

Varnes (1978) distinguished between lateral expansions: One type occurs in rock material and without the formation of an identifiable rupture surface, while the other corresponds to lateral expressions in cohesive soils that are underlain by materials that exhibit plastic flow (Fig. 2.8).

This process occurs when there is no obvious rupture. The movement has very different speeds depending on the type of material, being very slow in rocks due to their visco-plastic behaviour, and can be extremely fast in soils where collapse by liquefaction occurs (Zêzere, 2005), thus causing situations with serious consequences for infrastructures and populations.

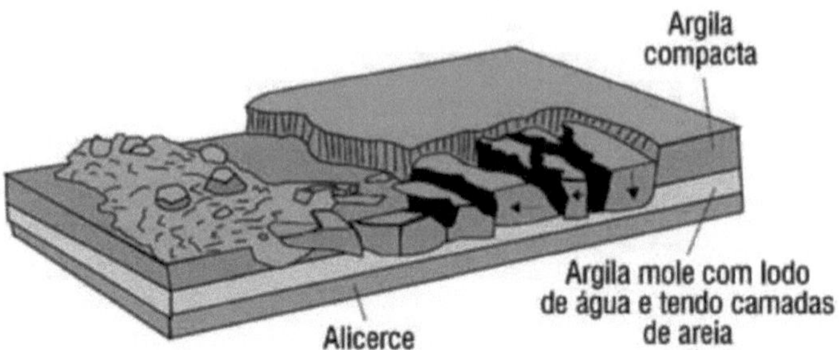

Figure 2.8) - Representation of lateral expansion showing the material that can undergo liquefaction, which is located below the surface layer, taken from Highland & Bobrowsky (2008).

2.2.5 - Flows

A flow is a continuous movement in which the shear surfaces are short-lived and often not preserved. It is a type of movement that can cover large areas and often occurs in soils, deposits, detrital material and rocks (Gerscovich D. M., 2009). The velocities of a flow correspond to those of a viscous fluid (WP/WLI, 1993), which are higher at the surface and decrease in depth.

The causes of flow movements are diverse and range from the action of gravity, temperature variation, the action of water, river dynamics, excavations, seismic action and increased load.

Rock flows are slow movements that occur in rock masses on the slopes of mountainous areas and which are highly diaclasticised or stratified (Zêzere, 2000).

Debris flows correspond to a mixture of fine (sand, silt and clay) and coarse (pebbles and blocks) heterogeneous materials, with a variable amount of water present, forming a mass that moves downwards, driven by the force of gravity and the sudden collapse of the materials (Zêzere, 1997).

Debris flow movements are related to periods of high rainfall. In this way, transitional terms can develop from floods, consisting almost exclusively of surface run-off water, to high density flows in which the suspended matter is considerable (Dias, 2006). These are usually movements that travel along existing channels and can reach considerable distances.

According to Varnes (1978) debris slides can turn into extremely fast debris flows or debris avalanches as the displaced material loses cohesion, increases its water content or encounters steeper slopes.

2.2.6 - Complex Movements

Complex movements correspond to the association of various types of instability movements in sequential phases (Andrade, 2008). They cover all situations in which, during their manifestation, there is a change in morphological, mechanical or causal characteristics.

The WP/WLI proposal (1993a, 1993b), followed by Dikau et al. (1996) and also by Cruden & Varnes (1996), advises using the term "complex movements" only for situations in which more than one type of movement can be distinguished along a temporal continuum (Zêzere, 2000).

Important causes of movement include the slope of the land and the action of water, the latter of which can cause saturation, often induced by heavy rainfall or changes in groundwater levels. Complex movements are also caused by anthropogenic factors such as the removal of vegetation on the slopes to build infrastructures or for agricultural and industrial production, waste production, water and sewage discharges, excavations and landfills.

2.3 - Factors related to slope instability processes

The stability of a slope is conditioned by geometric factors (height and slope), geological factors (related to the existence of planes and zones of weakness and anisotropy in the slope), hydrogeological factors (percolation and influence of water) and geotechnical factors or factors related to the geomechanical behaviour of the land (resistance and deformability) (Vallejo et al., 2002). The combination of the various factors mentioned above can define the conditions of rupture involving one or more surfaces. The possibility of failure and the mechanisms and models of slope instability are mainly influenced by geological and geometric factors.

Geological, hydrogeological and geotechnical factors correspond to conditioning factors and are intrinsic to natural materials (Table 2.4) (Vallejo et al., 2002). In addition to the factors that condition slope stability (known as passive factors), triggering or active factors cause failure, as they comprise a series of conditions. The latter correspond to external factors acting on soils or rock masses, altering their characteristics and properties, as well as the equilibrium conditions existing on the slope (Table 2.4).

The conditioning factors include: the lithological types and characteristics, the degree of alteration, the presence of water, the characteristics of the geological and geometric structures of the slopes, and the existence of vegetation.

The triggering factors also include human action, which generally contributes to the destruction of vegetation cover, the removal of land to open roads, the construction of buildings and the development of agriculture. Although this is not common in the study area, much of the literature considers the occurrence of earthquakes and vibrations, storms in coastal areas and temperature variations (contraction and dilation of rock materials) as triggering factors (Júnior & Longo, 2010).

Table 2.4 - Factors influencing slope instability movements, adapted from Vallejo et al. (2002).

Conditioning Factors (intrinsic to the materials or masses that make up the slopes)	Triggering factors (external factors that act on the materials or masses that make up the slopes)
- Stratigraphy and lithology.	- Static overloads.
- Geological structure.	- Dynamic loads.
- Hydrogeological conditions and hydrogeological behaviour of materials.	- Variation in hydrogeological conditions.
- Physical, resistance and deformation properties.	- Climatic factors.
- Natural stresses and stress-strain state.	- Variations in geometry. - Decrease in resistance parameter values.

CHAPTER 3

Characterisation of the study area

3.1 - Geographical framework

3.1.1 - Location and delimitation of the study area

The former colonial town and trading post of Catumbela, linked to the historical Portuguese presence in the Benguela region (Mombaka) since the beginning of the 17th century and, above all, to the strategic settlement next to the fertile agricultural lands of the Catumbela River delta, at the end of a route linking it to the hinterland, is today a flourishing town, where the proximity of the important socio-economic development axis of Lobito - Benguela, including the so-called Lobito corridor, is felt. As a result of this population and production growth, accentuated by the country's post-war economic development and recovery, Catumbela is also one of Angola's newest municipalities, and the most recent in Benguela province. As such, its surroundings were elevated to the category of municipality under Law 32/11 of 5 October 2011, with effect from 6 December 2011 (Figs. 3.1 and 3.2).

The privileged location of the Catumbela municipality on the centre-west coastline of Angola has also contributed significantly to this success. In fact, the municipality is positioned between the large demographic centres of Lobito and Benguela, both of which have many hundreds of thousands of inhabitants, playing, in a way, the role of a peripheral city to the former. Its territorial extension amounts to 814.71 km^2 and essentially corresponds to the space occupied by the former commune of Catumbela, to which was added the territory corresponding to the commune of Biópio. Defined in this way, the municipal area of Catumbela is bordered to the north by the municipality of Lobito, to the south by the municipality of Benguela, to the east by the municipality of Bocóio and to the west by the Atlantic Ocean (AMC, 2012).

As for the city of Catumbela, the geographical coordinates of its central area are 12° 25' 53" South latitude and 013° 32' 49" East longitude.

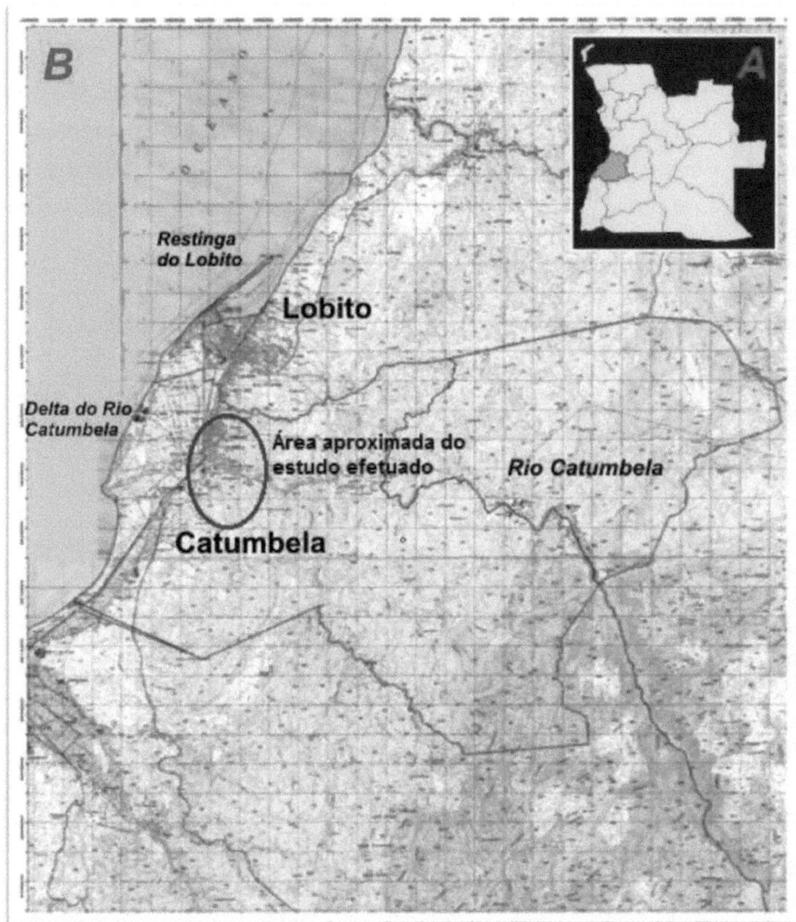

Figure 3.1 - A - Location of Benguela province in Angola, B - Location of Catumbela Municipality on the Topographic Map of Angola at a scale of 1:100,000, sheet 227/228 - Lobito. The red line represents the physical limits of the municipality and the blue ellipse the approximate limits of the town where the municipality is located.

The political and administrative division of Catumbela is relatively simple compared to other municipalities in Benguela province. It is made up of four communes, 12 villages and 53 neighbourhoods, home to an estimated population of 2,4375 inhabitants (Table 3.1).

Table 3.1. - Political-administrative division and population of the Municipality of Catumbela (AMC, 2014).

Commune	Area (Km)^2	Total Population	Population Distribution (%)
Bioscope	473.12	4038	1,79
Catumbela	80.31	134817	60,09
Gama	226.67	58489	26,07

| Baby Beach | 34.61 | 27031 | 12,05 |
| Total | 814.71 | 224375 | 100 |

The municipality's largest urban centre with the highest population density is Catumbela (Fig. 3.2), which has an estimated population of 1,348,17 inhabitants; at the opposite end is Biópio, with a population of 4038 inhabitants for an area of 473.12 km^2 (Table 1.1), according to data from AMC (2012). These records reflect the reality of the discrepancies between the coast, which is easily accessible, and the interior, which is still markedly rural and quite hilly.

Figure 3.2 - Vila da Catumbela (seat of the commune) - Catumbela Municipality.

3.1.2 - History

Catumbela was officially founded in 1836 (Bastos, 1912) following a decree issued by Queen Maria II of Portugal.

The previous name was Nova Asseiceira, in honour of the important victory of Pedro IV's troops over the absolutists, in one of the last clashes of the conflict, in the town of Asseiceira, in the region north of Lisbon.

The primitive colonisation site was located on the left bank of the Catumbela River (Bastos, 1912) and served as a springboard for the extension of Portuguese rule within the peripheral coastal area of Benguela, one of the oldest European settlement sites in Angolan territory since 1617.

Already at this time, on the right bank of the Catumbela River, there were some indigenous communities who mainly practised subsistence agriculture, growing maize, beans, sugar cane, pumpkin and sweet potatoes; raising a variety of livestock, including cattle, goats and pigs; and artisanal fishing.

The population was made up mostly of Mundombes, belonging to the "Herero" ethnolinguistic group. According to indigenous tradition, the settlement on the right bank of the Catumbela River was founded by a woman who was also its first soba. This population grew but was significantly reduced by an outbreak of smallpox, tuberculosis and sleeping sickness (Bastos, 1912).

During the first decade after its foundation, Catumbela was the centre of trade with the populations of inland Angola, particularly those from the central plateau, who came in search of coastal products (salt, dried fish, etc.) obtained from the European settlers.Due to the aggression suffered by the European settlers towards the local populations and others from Seles, Quissanje and elsewhere, in 1846 an expedition was organised to pacify the insubmissive "gentiles", which culminated in the construction of the Catumbela fortress, also known as the Reduto de S. Pedro (Fig. 3.3).

Figure 3.3 - The Catumbela fort (St Peter's Redoubt).

This fortification, erected on a site where slave traders used to trade pigeons, is now a monument of historical importance and there is still a commemorative engraving at the entrance that reads:

"It was done at the expense of the inhabitants of Bengella, to the honour of the worthy town council. The continuous insults made to the whites by the natives of this district led to the construction of this reducto. The garrisons of the brigs Mondego, Tamega and the corvette Relampago, commanded by Chief F. A. Glz. Cardozo, will submit them by order of the Governor General, the Honourable Pedro Alexandrino da Cunha. FX. Lopes Major do Ext°. traced and built 1ª stone 5 October 1846"[1]

After the pacification of the territory of Catumbela and the entire region, there followed a growth in the population and a period of development of economic activities, with the first colonial farms of S. Pedro, Lembeti and Maravilha do Cassequel springing up, where cotton and sugar cane were grown and used to produce brandy. The manufacture of roof tiles and bricks dates from this period, mainly for local consumption (Bastos, 1912).

The phase of Catumbela's prosperity came in the 1887s with the increase in the urban population, including Portuguese settlers, which corresponded with the construction of the first public infrastructure, including roads and bridges, telegraph and telephone services, basic sanitation, schools and a medical centre, churches, a town hall and a cemetery. The first bridge over the Catumbela River (1887) also dates from this period (Fig. 3.4).

The transition to the 20th century culminated with the implementation of the Benguela Railway, which quickly became the main communication route between Lobito and Benguela and between Lobito and Benguela and the interior of Angola. After the collapse of the first structure, the "D. Luís Filipe" bridge

(Fig.3.4) was built in 1905 by "Bridge & Roof Coy Ltd" of Glasgow, in a metal structure in front of the fort. Finally, in 1905, Catumbela was elevated to the category of town.

Figure 3.4 - Luís Filipe Bridge built in 1906 - Catumbela.

3.1.3 - Socio-economic characterisation

The dominant local ethnic group in Catumbela Municipality today is Umbundu and its people speak the Umbundu language in addition to Portuguese. Its main economic activities are mainly in the primary sector, namely agriculture, livestock and fishing. In the urban and neighbouring areas, commercial activities also occupy a significant proportion of the population. Artisanal sea fishing is the main activity of the populations settled closer to the coast, namely in the settlements of the Praia do Bebé Commune, which covers the whole of Catumbela's coastline.

At present, the municipality of Catumbela is considered to be the main industrial park in Benguela province, ranking third nationally.

Even in the recent past, before the expansion of the infrastructure of the port of Lobito, Catumbela was noted for its agricultural and commercial importance, including the production of sugar cane and palm oil, as well as the supply of food products to Benguela (Roquinaldo, 2013).

In the municipality's current area there is an industrial development centre where, in addition to a growing number of industries, the cement (Cimenfort) and beer (Cuca) factories, which are national ventures, also operate.

Also located in the municipality is Catumbela International Airport, inaugurated on 27 August 2012, which is also associated with the Lobito Corridor.

The new road bridge over the Catumbela River, called "4 de Abril", is one of the city's most recent tourist attractions. The structure is 438 metres long and includes a pair of access viaducts and a 170-metre platform over the river. It has two carriageways in each direction and pedestrian crossings.

The sea has great potential for tourism and is rich in a variety of fish, including horse mackerel, corvina and sardines, among other species of economic importance.

The Catumbela River reaches considerable flows for most of the year, which facilitates the practice of irrigated agriculture, the supply of water to the Municipality of Catumbela and other surrounding areas, as well as serving as a source of hydroelectric power.

3.1 . 4 - Climatology

The climate in the municipality of Catumbela is tropical and has a cooler season from 15 May to 15 August, with an average temperature of 20°C, and a hot season from September to May, where temperatures reach 32°C. However, previous studies, especially those carried out by Cruz (1940), show that Catumbela has a relatively constant local microclimate (average variation of less than 10 degrees).

The climate is hot and dry on the coast (despite the dryness, the relative humidity remains high throughout the year) and mesothermal on the inland sub plateau, with a moderately rainy water regime. The maximum temperature is 35.0°, the average 24.2° and the minimum 10.4°, with relative humidity of 79%, and very low average atmospheric rainfall, ranging from 50 to 250 mm (Cruz, 1940).

According to the Koppen classification (Feio, 1981) Catumbela falls into the types:

- BSh - Hot steppe climate of low latitude and altitude - Hot semi-arid climate, corresponding to the coastal strip of the municipality.

- Cwa - Humid temperate climate with dry winters and hot summers - Humid subtropical climate, which covers the interior of the municipality.

3.2 - Geomorphology

The morphostructural units of an essentially sedimentary nature, linked to the Meso-Cenozoic evolution of the continental margin of Angolan territory, within the framework of the episodes of rifting and distension of the South Atlantic Ocean, are traditionally subdivided into five sectors (Fig. 3.5), which are summarised as three coastal basins, known as the Congo Basin (located between the Zaire River and Ponta da Musserra), the Kwanza Basin (between Ponta da Musserra and the 12° 00' parallel) and, finally, the Namibe Basin (between the 13° 45'S parallel and beyond the country's southern border) (Torquato, 1978) and (Pinho & Carvalho, 2010).

In the following framework, the description of the geomorphology and stratigraphic units of Angolan territory was based on the Explanatory Note of the Geological Chart at a scale of 1:1,000,000 (Araújo & Guimarães, 1992). In terms of lithostratigraphy, the characterisation took into account the different eras (Archaic, Proterozoic and Phanerozoic) and periods recorded in the study area and its surroundings. For the magmatic, ultrametamorphic and metasomatic rocks, complexes dating from the Lower and Upper Archaic, Lower and Upper Proterozoic, Cretaceous, Cretaceous and Palaeogene, undifferentiated and of undefined age, were considered.

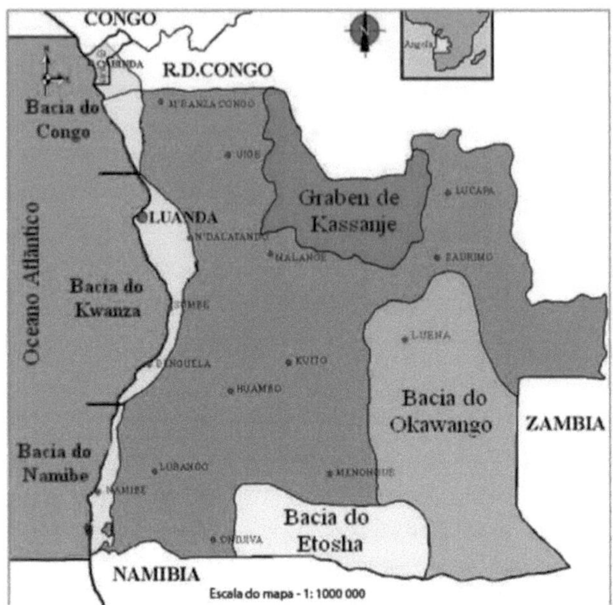

Figura 3.5   - Main sedimentary basins of the Atlantic margin (Araújo, et al., 1998;Victorino, 2012).

With regard to tectonics, the following major tectonic-structural elements have been defined: structures from the Lower Proterozoic, structures of the platform cover, tectono-magmatic activation zones of the platform and tectonic disturbances (DR, 2013).

Due to the specific characteristics of the relief, which is quite diverse as a whole, Angola's territory can be subdivided into two parts, called

Western and Eastern, respectively. Of these, the eastern part is more characterised by accumulation relief, while the western part is dominated by denudation relief, resulting mainly from intense Quaternary erosion phenomena (DR, 2013).

According to Honrado et al. (2010), these two major landscape domains of the Angolan relief are subdivided into the following zones (Fig. 3.6; the numbering below follows the legend in this figure):

I Western Part: 1 - Central Plateau; 2 - Angola's marginal mountain range; 3 - The Maiombe Plain with a slightly hilly terrain; 4 - The Zenza - Loge mountain range; 5 - The undulating Kwanza - Longe Plain; 6 - The strongly dissected Cuango Plain; 7 - The Cassanje Depression; 8 - The Coastal Depression.

II Eastern part: 9 - Lunda plateau: 10 - Eastern plain; 11 - Cunene proluvial plain: 12 - Cameia - Lumbate depression; 13 - Upper Zambezi ridge (Other conventions: 14 - the most important steps formed by tectonic and denudation effects; 15 - boundary between the eastern and western parts).

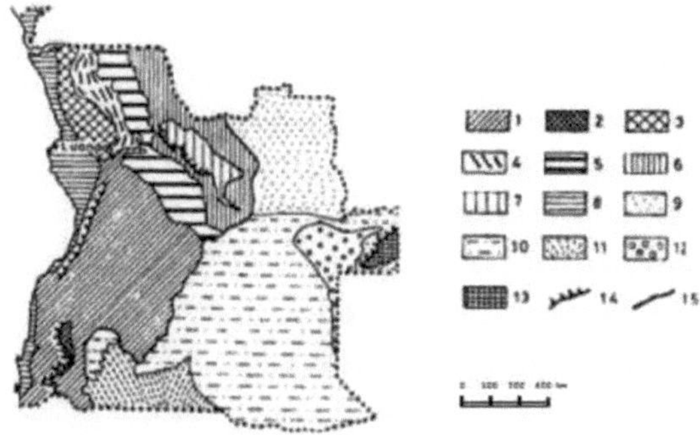

Figura 3.6   - Outline of the geomorphological units of Angola's territory, adapted from Honrado et al. (2010).

Assuming a genesis subordinated to this dynamic and due to its proximity to the coast, it is possible to distinguish, in the area of the municipality of Catumbela, three large geomorphological units of regional scope, respectively called: (1) coastal strip, (2) transition zone and (3) marginal mountain chain.

The coastal strip is developed on a substrate of detrital, mixed and carbonate sedimentary rocks, part of the Meso-Cenozoic onshore Benguela Basin. It also includes a vast flat area of alluvial plain and coastline, related to the development of the Catumbela River delta. This unit is made up of sedimentary geological materials, mainly of marine origin, covered by marine sands on the beaches and by alluvial materials along the banks of the rivers and lowlands. It has the least rugged relief of all, covering lower elevations. It includes a variety of forms related to the structural evolution of the Atlantic margin and differential erosion in calcareous-marl and detrital units, as well as the Pleistocene embedment of drainage networks, marine abrasion at different altitudes and Holocene accretion. These include, for example, coastlines, dambas, fault scarps, abrasion platforms and palaeoarribes (Carvalho, 1963).

The Transition Zone, located to the east of the previous one, represents the first pre-Mesozoic basement unit in the Benguela Basin. It consists of a strip roughly parallel to the previous one, with a width of between 15 and 20 kilometres. It is formed on a depressed surface ("baixa dos gnaisses") and is made up of rocks from the metamorphic complex in which gneisses, migmatites, mica schists and volcanic rocks similar to dolerite predominate (Galvão & Silva, 1972).

The marginal mountain range, located further inland, establishes the transition to the plateau regions of Angola's central interior. It is mainly made up of eruptive rocks. It is characterised by a rejuvenation of the relief and it is in this area that the highest altitudes in the whole region are found. The dominant rock type is medium-grained granite (Galvão & Silva, 1972).

A description of the geomorphology, geology and tectonics of Catumbela can be found in the Notícia Explanativa da Carta Geológica n° 227-228 Lobito, at a scale of 1:100,000 (Galvão & Silva, 1972).

Studies carried out in the region in question have revealed that it is fundamentally home to marly limestones, dolomitic limestones and oolitic-oncolithic limestones, forming successions of strata that are weakly tilted to the west, but crossed by a network of fractures which have resulted in local deformations, including folding (Pereira, 2001). At the base is an evaporite complex, essentially made up of gypsum. In the southern part there are also sandy-clay sediments related to the Holocene sedimentary deposition of the Catumbela River, in an extensive deltaic plain. The geomorphological characteristics of the region are the result of the strong contrast in resistance between the limestone formations that make up the limestone massif and the sedimentary deposits of the alluvial, deltaic and coastal plains.

In the limestone massif, the slopes tend to have a resistant limestone cornice at the top, which is steeply sloping, and in the middle and lower parts a concave slope with a variable gradient (usually between 10 and 20 per cent). These include the "dambas", where the slopes tend to be steep and more or less verticalised. In these contexts, calcareous, heterogeneous and relatively consolidated slope deposits, sometimes with some traces of stratiform organisation within them, are particularly common. They vary greatly in thickness, reaching more than a dozen metres in height, and are mainly made up of limestone debris of thermoclastic origin, embedded in a clay matrix (Pereira, 2001).

The most modern forms of accumulation are the most interesting for understanding recent geomorphological evolution. According to Carvalho (1963), this evolution can be summarised as follows:

1 - During the Flandrian transgression, in the course of the Holocene, the Catumbela River delta would have formed, over which, in a regressive phase, this watercourse would have shifted its bed from a position further north and orientated towards the current Lobito sandbank, until it occupied its present location (Fig. 3.7).

2 - During the regressive phase that followed the maximum transgression, the sandy deposits of the strip behind the current sandy beaches would have accumulated. This corresponds to a regressive phase at the end of the Flandrian.

3 - Today, the regressive behaviour of this sector of the coast is destroying those deposits, at least locally (Galvão & Silva, 1972).

Fig. 3.7 - Aerial panorama of the lower course and delta of the Catumbela River, during a flood episode (Image composed from Google Earth, (2014).

The terrain in the coastal area of Catumbela is characterised by the presence of areas of fairly low elevation, of coastal or alluvial plains subject to flooding episodes, which are associated with coastal or alluvial processes. There are also areas of steeper relief, as mentioned above, cut by networks of deep valleys and vertical flanks (dambas), for example in the neighbourhoods of Tata, Cambambe, Chiúle, etc. In this region, according to Feio (1960), coastal lows have formed where the land is flattened and has elevations of only a few metres above sea level, and whose sedimentation processes are of upper Pleistocene and Holocene age. Inland, there are also two levels of marine terraces that correspond to ancient platforms of deposition and abrasion. In turn, the alluvial plain of the Catumbela River, whose elevations are around 3 to 6 metres, is made up of sandy-clay deposits such as reddish sands with calcareous concretions, light sands, calcareous sands and alluvial sediments (Feio, 1960).

3.3 - Geology

3.3.1 - Regional Geology

The geology of Benguela Province comprises a succession of units of Precambrian to Cenozoic age (Figs. 3.8 and 3.9), of which the sedimentary ones are centred on the coastal region of the Atlantic coast, within the context of the Benguela Coastal Sedimentary Basin, of Cretaceous to Quaternary age. The latter is precisely the context of the study area, located in the municipality of Catumbela, where marine carbonate rocks and cover sediments predominate.

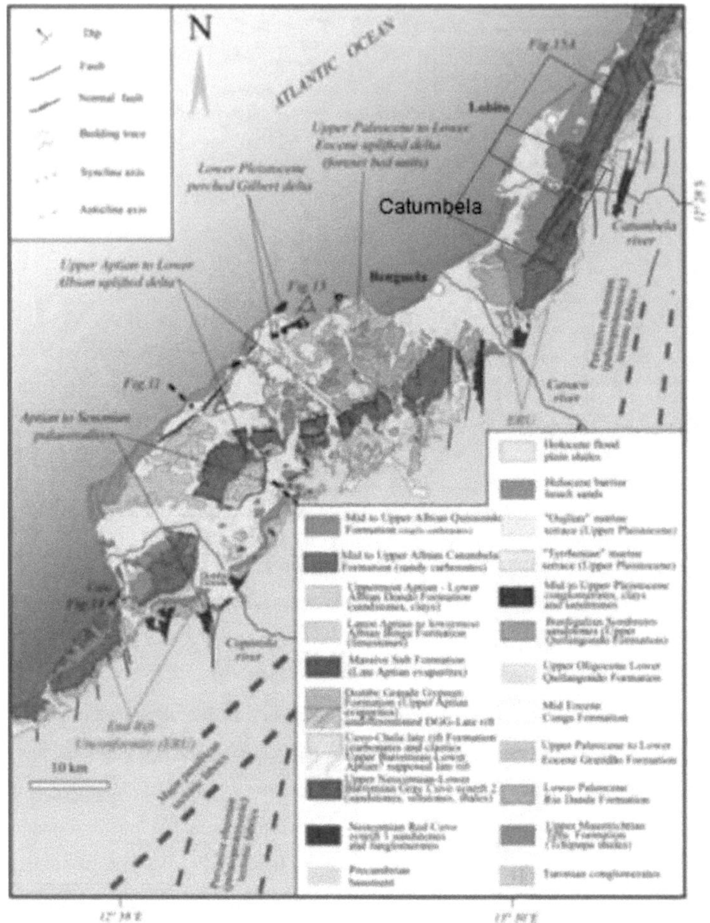

Figure 3.8 - Geological map of the Benguela basin adapted from Guiraud et al. 2010.

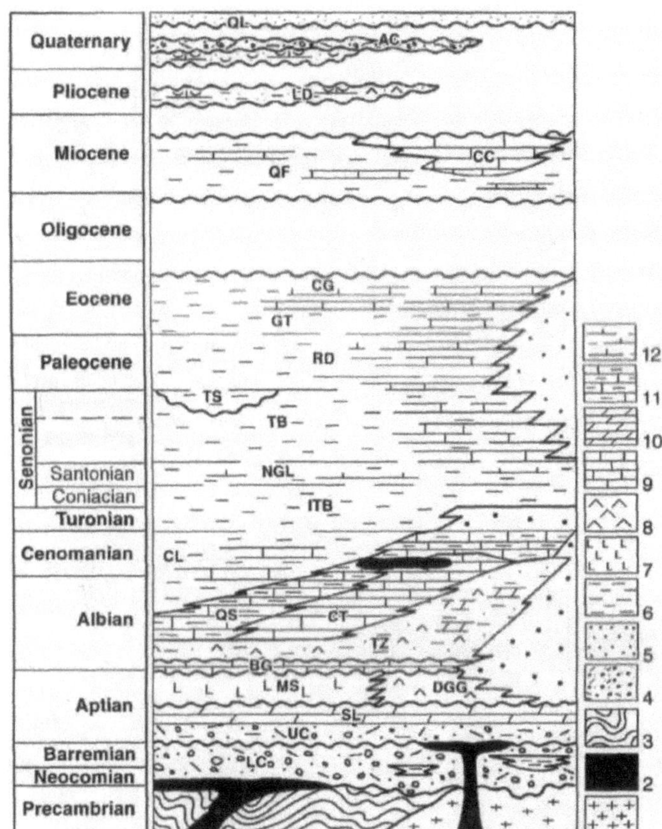

Figura 3.9   - Synthesis of the stratigraphy of the Kwanza Basin, including the Benguela Basin in its southern sector, (GeoLuanda 2000) *apud* (Victorino, 2012).

<u>Key:</u> 1 - Intrusive rocks, granites; 2 - Extrusive rocks, basalt; 3 - Metamorphic rocks; 4 - Conglomerates; 5 - Sands; 6 - Clay shales; 7 - Evaporites; 8 - Gypsum; 9 - Carbonates; 10 - Carbonates and dolomites; 11 - Calcilutites; 12 - Marls.

AC - Grey Sand Formation; **BG** - Binga Formation; **CC** - Cacuaco Formation; **CL** - Cabo Ledo Formation; **CG** - Cunga Formation; **CT** - Catumbela Formation; **DGG** - Dombe Grande Formation; **ITB** - Itombe Formation; **GT** - Gratidão Formation; **LC** - Lower Cuvo Formation; **LD** - Luanda Formation; **MS** - Massive Salt Formation; **NGL** - N'Golome Formation; **QF** - Quifangondo Formation; **QL** - Quelo Formation; **QS** - Quissonde Formation; **RD** - Rio Dande Formation; **SL** - Chela Formation; **TB** - Teba Formation; **TS** Tchipupa Shales; **TZ** - Tueza Formation; **UC** - Upper Cuvo Formation.

Considering Sheets 227/228 of the Geological Chart of Angola (scale 1:100,000) (Fig. 3.10), which shows the Lobito region, there are three distinct geological zones, orientated roughly NNE-OSO, which are clearly related to the morphological domains described above:

The first is made up of a strip stretching from north to south of the mapped area, close to the coast, formed by sedimentary rocks whose ages cover the Neocomian to Recent interval (Fig. 3.9).

A second zone is called the "Metamorphic Complex" and occupies a structurally depressed area, conditioned by NNE-OSO faults and situated between the granitoid rocks of the eastern strip of the chart and the Meso Cenozoic sedimentary border. The main lithotypes present are gneisses, granite-gneisses and mica schists, and have been considered to be of older age within the regional geology, going back to the Precambrian. The mapped area corresponding to the Metamorphic Complex is also cut by several dykes of diabase intrusive rocks, varying in length and predominantly in a NE-SO direction. Other intrusive bodies present in some places have a composition that varies between gabbro and diorite. The age of these bodies still needs radiometric studies, but their relationship with rifting and the beginning of the structuring of the Atlantic margin seems certain, especially from the Neocomian onwards.Finally, the third mapped area is dominated by extensive outcrops of granitoid rocks, corresponding to the mapped spots in the eastern strip of the chart. The ages of these bodies range from the base of the Aptian to the age of the rocks of the Metamorphic Complex (Galvão & Silva, 1972).

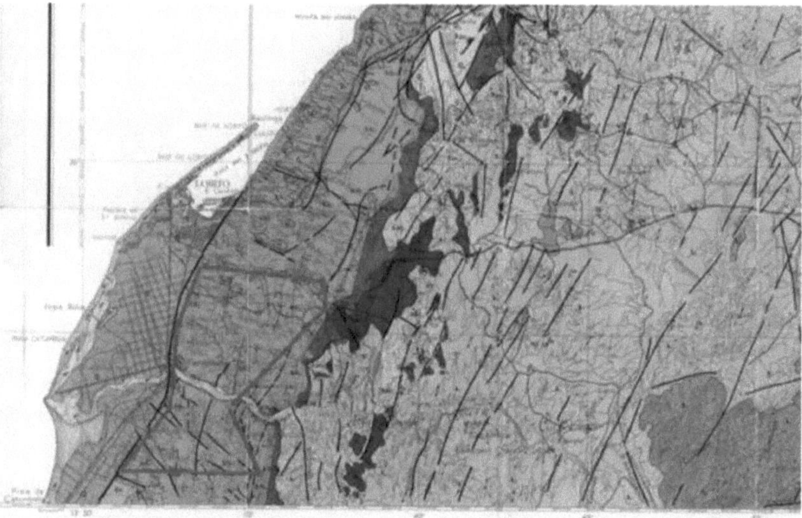

Figura 3.10   - Location of the study area based on the Geological Map of Angola sheet 227/228 - Lobito, within the upper Albian unit (Alb 3).

3.3.2 - Local geology

The geology of the easternmost areas of the Municipality of Catumbela comprises a socle that includes very heterogeneous units of magmatic and metamorphic rocks, corresponding to the two zones described above (Fig. 3.9). The thick sedimentary cover, essentially carbonate, of the Benguela Basin rests discordantly on top of these and as you move towards the coast, especially the Catumbela and Quissonde formations (Tavares, 2005), whose strata of limestone, marl limestone and marl outcrop extensively on the slopes of the Catumbela River valley and constitute the geological substrate of the municipality's headquarters and the hills bordering the deltaic plain.

There are also various deposits of marine and fluvial terraces, given that Catumbela falls within the sedimentary belt of Lobito and Benguela, where at least two altimetric levels can be defined, evidenced by fossil cliffs and platforms developed on Cretaceous limestones, with or without sandy cover, as well as deposits laid out along the Lobito-Benguela road (Carvalho, 1957). However, the thickest sedimentary cover, mainly of a sandy and sandy-gravelly nature, is found on the deltaic plain bordering the municipal centre. The lithological composition of the materials that make it up is, of course, due to the geological nature of the areas located further upstream, within the drainage basin of the Catumbela River.

Also important from a geological and economic point of view, the main possible occurrences in the region are (Governo Provincial de Benguela, s/d):

> * Industrial minerals:
>
>> Sands, clays, gypsum, malachite and copper;
>
> * Ornamental stones:
>
>> Granites, limestones.
>
> * Cladding stones:
>
>> Sandstones, Gneisses, Quartzites.
>
> * Industrial rock:
>
>> Granites and limestones.

3.3.2.1 - Local geological units

An analysis of the geological map for sheets 227/228 - Lobito (scale 1:100,000) (Galvão & Silva, 1972) reveals the existence of several units:

a) Holocene and Pleistocene deposits

They correspond to cover deposits of Pleistocene and Holocene age, made up of sedimentary materials, almost always unconsolidated, deposited or in transit. The mapped area contains the units designated **a'**, **a''**, **Q, Qi and Q2**, which respectively designate beach sands, alluvial fans, miscellaneous deposits (tuffs, limestones, sands, gravels and clayey sediments), terrace deposits (high terraces - elevation > 40 metres) and clayey-sandy coastal plain sediments. In particular, a distinction should be made:

Beach sands (a'), present on the current coastline and developed in front of sandy platforms, whose deposits represent the *backshore* and to which Carvalho (1961) attributed a Flandrian age.

*Modern alluvial fans (*a''*) which, in* general, are made up of fine, slightly clayey sands containing organic matter and covered by surface gravels of materials in transit, especially during periods of flooding.

Sediments from the deltaic plain of the Catumbela River (Q), spread over the extensive coastal strip between Lobito and Benguela, and forming a platform at heights of between 3 and 6 metres, made up of two dominant types of sediment: dark silt and more or less muddy sands; beach sands of a more recent age than the previous ones.

Terrace deposits (Qi) linked to the Quaternary evolution of the coastline and the development of raised beaches, on which Feio (1960), in a detailed study carried out in the Lobito region, noted the presence

27

of the following altimetric plateaus with Pleistocene levels:

175 metre platform - Calabrian

93 metre platform - Siciliano I

Levels 46 - 50 metres - Siciliano II

28 metre level - Tyrrhenian I

13 metre levels - Tyrrhenian II

These deposits can be seen in various places such as the Acongo, Nangolo and Morro do Galo neighbourhoods and may correspond to the 175 metre platform with Calabrian age (Feio, 1960).

b) Pleistocene deposits

The consolidated deposits of coarse, feldspathic sandstone, sometimes containing pebbles and fossils of archaeid bivalve molluscs (Senilia senilis Linné, 1758), which are found in small patches on the Cretaceous surface at altitudes varying between 100 and 135 metres (Galvão & Silva, 1972), are probably of this age.

These deposits extend further south, in the region between Negrão and Luongo and even near Gama and Vimbalambi. Due to their characteristics, these flaps should be linked to the "Sombreiro Sandstones" (Benguela), a discordant unit over the Quifangondo Formation, of Miocene age (Galvão & Silva, 1972).

c) Cretaceous formations

According to the nomenclature used in the explanatory note to the geological map of Lobito, which we have been following, the stratigraphic units representing this period in the Catumbela municipality are called:

c.1 - Albiano

c.1.1 - Upper Albian (Alba)

It forms the base of the Cenomanian. It crosses the entire coastline to the east of the coastal plain sediments (Q). It is made up of impure limestone, marls, sometimes nodular, calcareous clays and lutites (Galvão & Silva, 1972). In the study area it is thick with conglomeratic characteristics, especially in the Luongo region, with ammonites of the genera *Pervinquiena, Mortonicerase* and *Elobiceras, indicative of this age.*

c.1.2 - Middle Albian - Lower Albian (Alb2)

This unit includes the area that stretches from the north to the south of the Charter, irregularly parallel to the coast and which is limited to the west by the limestone-marl formations of the upper Albian/Cenomanian and to the east by the sandstone formations, lagoon-continental formations of the lower Albian and lower Aptian formations (Fig. 3.10) (Galvão &

Silva, 1972). Equivalent to the Quissonde Formation sensu Tavares (2005) and later authors, the unit represents the outermost sectors of the carbonate ramp recorded onshore. It consists of fairly homogeneous, rhythmic sequences made up of white or greyish, compact, sub-crystalline, massive or stratified in thick layers, oolitic or pisolitic limestones, including hyaline quartz grains. These limestones can be seen in the Luongo and Pedreira regions. Studies have shown them to be remarkably uniform in terms of composition, and the percentage of carbonate fraction is generally very high, reaching 97% (Galvão & Silva, 1972).

c.1.3 - Lower Albian (lagoon-continental formation) - (Albi Ig)

This equivalent unit is equivalent to the Catumbela and Tuenza Formations sensu Tavares (2005) and later authors, and is representative of the innermost palaeogeographical sectors of the Benguela Basin. It contacts the limestone formations of the middle Albian to the west and the formations of the lower Aptian, upper Aptian and base complex to the east (Fig. 3.10) (Galvão & Silva, 1972).

In the study area, it crosses the Catumbela hill region, close to the current Pedreira neighbourhood, and advances into the Gama region. It is made up of limestones in thick, massive benches, stony limestones and different types of siliciclastic lithologies that reflect the influx of sediments into the carbonate ramp.

c.2 - Aptian

c.2.1 - Upper Aptian - (Apc2)

It consists of marly limestones with fossils of Pholadomya pleuromyaeformis Choffat and Natica feioi Choffat, as well as sublithographic limestones, lagoon limestones, limestones with algae, dolomitic limestones with gypsiferous marls, feldspathic sandstones and limestones in platelets (Galvão & Silva, 1972). They are equivalent to the Bingo Formation sensu Tavares (2005) and later authors, indicating a gradual transgressive evolution within the onshore Benguela Basin.

It is bordered to the west by limestones from the middle Albian and upper Aptian, as well as by lagoon-continental sandstone formations from the lower Albian. To the east, this unit comes into contact with rock formations from the "Metamorphic Complex" (Ai) and limestone rocks from the upper, lower and middle Aptian (Fig. 3.10).

c.2. 2 - Middle - lower Aptian - (Apci)

This unit, equivalent to the Sal Massivo Formation sensu Tavares (2005) and later authors, is limited to the west by limestones from the middle Albian and upper Aptian and by

lagoon-continental sandstone formations of the lower Albian and to the east by rocks of the Metamorphic Complex and also by limestone formations of the upper Aptian. It varies greatly in thickness and consists mainly of evaporite rocks rich in gypsum, whose genesis is related to deposition in a lagoon environment and subsequent diapiric activity.

c.2.3 - Pre-Aptian or Lower Aptian - (ci)

This continental unit is always in contact with the Metamorphic Complex or relatively close to it. It is visible in the Biópio, Central da Cassequel and Gama regions.

It is made up of thick greso-conglomeratic levels containing abundant clasts of basement rocks. They represent the oldest unit of the Cretaceous (Galvão & Silva, 1972) and are equivalent to the Cuvo Formation sensu Tavares (2005) and later authors.

3.4 - Soils

Soils have minerals that result from the weathering of a parent material (rock) by mechanical disintegration or chemical decomposition and whose transformation takes place in a given relief and climate over time (Caputo, 1988). The main types of soil include:

a) *Residual* Soils - are those that remain in the place of the original rock, with a gradual transition from the soil to the rock.

b) *Sedimentary soils* - these are *soils that* have suffered the action of transport agents and can be alluvial (transported by water), aeolian (by wind), colluvial (by the action of gravity) and glacial (by glaciers).

c) *Soils of Organic Formation* - these are *soils* of essentially organic origin, whether vegetable (plants, roots) or animal (shells).

The characteristic soils of Catumbela are essentially sedimentary: mostly alluvial soils, limestones, clays and tropical fersialitic soils (luvissolos, nitissolos) and show indicators of variable fertility with certain mineral reserves on the coast (Santos & Zacarias, 2010). Soils with good physical and chemical characteristics abound in the interior of the region, such as the so-called tropical arid soils, alluvial soils (fluvissolos), calsialitic soils (mostly luvissolos), and also calcissolos.

The alluvial plain of the Catumbela River is irrigated by small diversions from the river and is an agricultural area with good natural conditions (Feio, 1960). Part of this land has now been converted into land reserves for industry and agriculture. In this perimeter, 18 series of non-salty soils and 9 salty soils have been defined and characterised. These include the fine and medium to coarse-textured Catumbela alluviosols; fine-textured limestone pore coluvissols; coastal sandstones with and without limestone influence.

CHAPTER 4

Description of the Methodology Adopted

The methodology adopted in this work proved useful in identifying and defining the different characteristics of the slopes, as well as the instability situations that occur on the slopes studied and which are located in the municipality of Catumbela. It was found that the factors associated with instability movements are related to the geotechnical characteristics of the land, the conditions of its occupation and the type and quality of construction.

According to the methodology used for the slopes studied and taking into account the different parameters of the worksheet adopted, the geometric components of the slopes, the lithological types, the discontinuities, the existing vegetation, the types of instabilities occurring and their velocity values, the degree of instability activity and its volume, the consequences and the internal and external causes were characterised.

This study was based on a methodology that encompassed five distinct phases (Fig. 4.1). The first phase consisted of collecting and interpreting relevant literature for the development of the work, where information was sought on the geographical, historical and economic aspects of Benguela Province and the study area, while also seeking to systematise the various contributions of authors in the area of instability movements; also in the first phase, information was collected from the Catumbela Municipal Administration, the Provincial Directorate of the Institute of Geodesy and Cartography of Angola (IGCA), where the cartographic map of the region was obtained, information was requested from the Provincial Directorate of Geology and Mines, regarding the geological maps of the region (without success), and from the Provincial Directorate of Spatial Planning, Urbanism, Housing and Environment (DPOTUHA), through which land use maps were obtained.

In the second phase, study sites were chosen in the Catumbela area, and a worksheet was drawn up on instabilities (Table 4.1) and its most important parameters, such as the geometric aspects of the slopes, the lithological types present, the existing vegetation, the structural aspects, the types of instability, the existence of works to stabilise the slopes, the speed of the instability movements, the characterisation of the states of activity, the dimensions of the instability, the consequences and the external and internal causes of the instabilities, as well as a preliminary diagnosis of the current and past instability conditions. To identify areas of instability, topographical and geological maps on a scale of 1: 100 000 were used, among others. The Rockfall Hazard Rating System (RHRS) was used and the parameters of this classification were analysed. Linear sampling was used to characterise geological-geomechanical aspects, such as the spacing and opening of discontinuities, the fill material and the roughness of the discontinuity surfaces. In some slopes, the Schmidt hammer was used to determine the hardness of the lithological materials present.

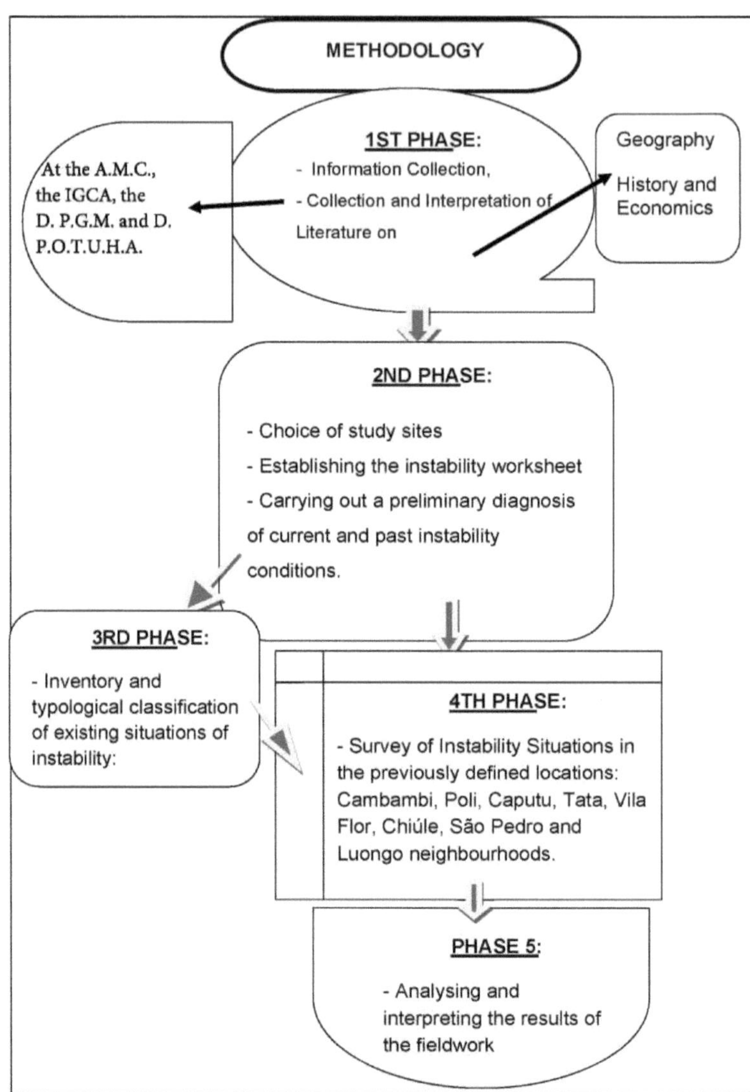

Figure 4.1 - Flowchart of the Methodology adopted.

Table 4.1 - Parameters in the worksheet on unstable situations.

1 - Name of slope/verge:	Date:
2 - Location:	
3 - Length (in metres) of the slope/verge:	
4 - Height (in metres) of the slope/verge:	
5 - Slope/verge inclination:	
6 - Lithological types:	
7 - Vegetation present:	
8 - Sketch of the lithology and geological structures (interpretive cross-sections and/or interpretive front of the slope/verge):	
9 - Types of instability:	
10 - Stabilisation work: Indication of the presence of stabilisation or protection work on the slope/verge; if present,	

indicate which work has been carried out.
11 - Sketch of the instability (you must make an interpretative drawing with a scale of the instability and its position on the slope/verge):
12 - Velocity of slope movements (must be filled in if there are movements).
13 - States of Instability activity (Adapted from Unesco WP/WLI (1993)):
14 - Dimensions of instability
15 - Consequences of instability
16 - External causes of instability
17 - Internal causes of instability

The third phase of the study included an inventory and typological classification of instability situations in the region and an analysis of their relationship with the lithology, morphology, geological structure and geotechnical characteristics of the formations affected. In the third phase, the recommendations and suggestions provided by the scientific advisors were followed, with special emphasis on defining the various types of instability, their mechanisms, causes and consequences, and a study of the geological characteristics of the study sites was also carried out.

In the fourth phase, a detailed survey of instability situations was carried out at the sites previously defined in the previous phases, in the neighbourhoods of Cambambi, Poli, Caputu, Tata, Vila Flor, Chiúle, São Pedro and Luongo. In the case of slopes located along roads, use was made of the Rockfall Hazard Rating System (RHRS) and the definition of its various parameters.

The fifth phase involved analysing and interpreting the results obtained during the fieldwork, specifically the main parameters that were considered in the worksheet, especially the size of the slopes, the lithology and vegetation present on the slopes, the types of instability and the internal and external causes of these instabilities, as well as their main consequences at the study sites. The research made it possible to identify the main similarities and differences in the occurrences of instability on the slopes studied. The use of the RHRS classification made it possible to define the possibility of instability movements on road embankments. In the end, some considerations were made about the main consequences of instability situations, and recommendations and actions were drawn up regarding the mitigation and prevention of instability occurrences, in order to allow for better management and planning of the territory and enabling better well-being for the population.

4.1 - Description of the Rockfall Hazard Rating System (RHRS)

In this study, this method contributed to assessing the risk related to slope instability situations along motorways. There have been instances of falling blocks on the slopes studied, which can cause damage or even casualties to road users. The Rockfall Hazard Rating System (RHRS), according to Pierson et al. (1990), is a method for analysing the risk of excavation/roadway slopes where blocks of rock fall.

The method is a first approach to analysing the risk of instability, particularly rockfall. The classification makes it possible to prioritise the interventions to be carried out or the need for more in-depth studies of instability situations.

This system has been used by various Transport Departments in the USA, and in some countries with adaptations - Russel et al. (2008), and also in other countries such as Italy (Buddeta & Panico, 2002); Li et al., (2009); in Spain with Luciano (2008) and in Portugal by Gomes (2009), in the characterisation of instability in railway embankments in central Portugal, and by Pires et al. (2012) in the study of embankments on the

EN353 in Idanha-a-Nova.

The RHRS classification uses parameters related to the amount of traffic, traffic speed, driver visibility, road dimensions, ditch effectiveness and geological characteristics, as can be seen in Table 4.2.

Table 4.2 - Parameters considered in the Rockfall Hazard Rating System (Adapted from Hoek, 2007)

CATEGORY			Classification and Scoring Criteria			
			3 Points	9 Points	27 Points	81 Points
SLOPE HEIGHT			7,62m	15,24m	22,86m	30,48m
DITCH EFFICIENCY			Good retention capacity	Moderate retention	Limited Retention	No Holds Barred
MEDIUM RISK FOR VEHICLES			25% of the time	50% of the time	75% of the time	100% of the time
DECISION VISIBILITY DISTANCE			Adequate viewing distance (100%)	Moderate viewing distance (80%)	Limited viewing distance (60%)	Very limited viewing distance (40%)
TRACK PLATFORM WIDTH			13,41m	10,97m	8,53m	6,10m
GEOLOGICAL NATURE	CASE 1	CONDITION STRUCTURAL	Discontinuous diacrasis, favourable orientation	Discontinuous diacrasis, random orientation.	Discontinuous diacrasis, Unfavourable orientation	Continuous diacrasis, unfavourable orientation
		ROCK FRICTION	Rough, irregular	Wavy	Smooth	Filling clay or polished
	CASE 2	CONDITION STRUCTURAL	Some distinct erosion characteristics	Occasional erosion features	Many erosion features	Huge erosion features
		DIFFERENCE IN EROSION RATES	Small difference	Moderate difference	High difference	Extreme difference
BLOCK SIZE			0,31m	0,61 m	0,91m	1,22m
NUMBER OF BLOCK FALLS PER EVENT			2,29m³	4,59m³	6,88m³	9,18m³
PRESENCE OF WATER ON THE SLOPE			Low to moderate rainfall; no periods of freezing without the presence of water on the slope	Moderate rainfall or short periods of freezing or intermittent presence of water on the slope	High rainfall or long periods of freezing or continuous presence of water on the slope	High rainfall and long periods of ice, or continuous presence of water on the slope and long periods of freezing.
HISTORY OF FALLING BLOCKS			Few falls	Occasional falls	Frequent falls	Frequent falls

Each of the parameters has been assigned a weight ranging from 3 to 81 points (Table 4.2). For each case study, all the parameter values must be added together. The lower the score, the lower the risk of instability. When the sum of the weighted values is less than 300, the risk is considered to be low and stabilisation measures should be taken within a medium timeframe, while when the sum of the values is greater than 500, the risk is considered to be high and urgent measures should be taken.

4.1.1 - Slope height

This parameter corresponds to the value of the vertical measurement of the height of the slope. It is an important parameter in analysing the risk of falling blocks, as it is related to the potential energy with which unstable materials hit vehicles or passers-by on roads, so the greater the height, the higher the score for this parameter.

4.1.2 - Ditch effectiveness

The parameter corresponds to the retention capacity of the ditch in relation to the instabilised rock material, in order to prevent it from reaching the roadway. The following characteristics must be taken into account when defining the values for this parameter:

- Height and slope gradient.
- Ditch dimensions.
- Dimensions of the rock blocks.
- Presence of irregularities on the surface of the slope.

4.1.3 - Average Vehicle Risk (AVR)

The average risk for vehicles is a parameter that determines the percentage of risk for a given vehicle on a road where rockfalls can occur. It is a parameter determined from the average hourly traffic intensity (AHT), the length of the section and the speed limit on the road in question, as shown in equation (4.1). The lower the speed limit and the higher the road traffic, the greater the likelihood of a vehicle being hit by falling rock.

$$AVR = \frac{TMH \times Length\ of\ Section\ (Km)}{Speed\ limit\ (km/h)} \times 100 \qquad (4.1)$$

4.1.4 - Decision visibility distance (DVD)

It is a parameter related to the distance that is reserved for a driver in order to manoeuvre or make an instant decision in relation to the presence of unstable rock material on the roadway, thus avoiding an accident.

The decision visibility distance (DVD) parameter corresponds to the ratio (equation 4.2), in percentage terms, between the local visibility distance (DVL) and the decision visibility distance (DD), the latter being related to the speed at which vehicles travel, the friction between the tyres and the road surface, as well as the gradient of the roadway.

$$DVD = \frac{Real\ DVD}{DVD\ proj} \times 100 \qquad (4.2)$$

4.1.5 - Track width including paved verges

This parameter corresponds to the manoeuvring space that a given driver can have on a road in order to avoid colliding with unstable blocks.

It is determined perpendicular to the direction of the carriageway, including paved verges.

Thus, if the carriageway is wide, the risk related to blocks coming from the slope will be lower; whereas for narrower carriageways, the risk is higher and consequently the weight values are higher.

4.1.6 - Geological characteristics

Geological characteristics are a parameter of the RHRS classification that is determined differently from the previous ones, as it considers two different cases, either of which can be chosen (Table 4.2):

- Case 1 is considered for slopes in which discontinuities such as diaclases, faults and stratification planes are the predominant structural characteristics that condition possible ruptures.

- Case 2 is related to slopes where instability situations, such as falling blocks of rock, are associated with erosion processes or excessive slope.

4.1.7 - Size of blocks or number of blocks resulting from rupture per event.

The parameter comprises two possibilities: to take into account the volume of individual blocks or to consider that several blocks may instabilise in a given event. The choice should take into account the type of instability that is most likely to occur. Maintenance records should be used to define this parameter or existing situations should be analysed. The greater the volume of unstable rock material, the greater the risk associated with these instabilities.

4.1.8 - Climate and presence of water

Water is an instabilising element in rock masses, and the most important of these are the percolation of water, the increase in pressure caused by the presence of water and the actions related to freeze-thaw.

To define the weighted value of the parameter, the amount of annual rainfall at the site must be taken into account. If the annual rainfall is less than 508 mm, the slope site is in a low rainfall zone. If the annual rainfall is more than 1270 mm, then the slope is located in a high rainfall zone. It is also important for this parameter to define the history of block falls, if any, during periods of heavier rainfall.

4.1.9 - History of block falls

In this parameter, historical records of rockfall volumes over the years should be analysed in order to make predictions of possible instabilities and their consequences.

The frequency and size of rockfalls should be determined, as well as the intensity with which they have hit people or property. Recording this information

This information also has added value for a statistical analysis of the consequences of instability situations.

The description of the method takes into account the original requirements of the RHRS classification, it represents a simple and inexpensive methodology, field work does not require cutting or loading equipment, and it should be noted that the parameters can be defined from determinations at the study site.

4.2 - Determining hardness using the Schmidt hammer

The Schmidt hammer (Fig. 4.2) is a portable, relatively inexpensive instrument that can be used to determine the surface hardness of a rock mass. The test can be carried out quickly and efficiently, both in the laboratory and on site. The hammer has a metal rod at one end, which is pressed against the surface to be tested, allowing the Schmidt hardness (R), whose values range from 0 to 100, to be determined by the bounce of a spring located inside the Schmidt hammer. The surface to be tested should preferably be slightly altered and can be cleaned before the tests are carried out.

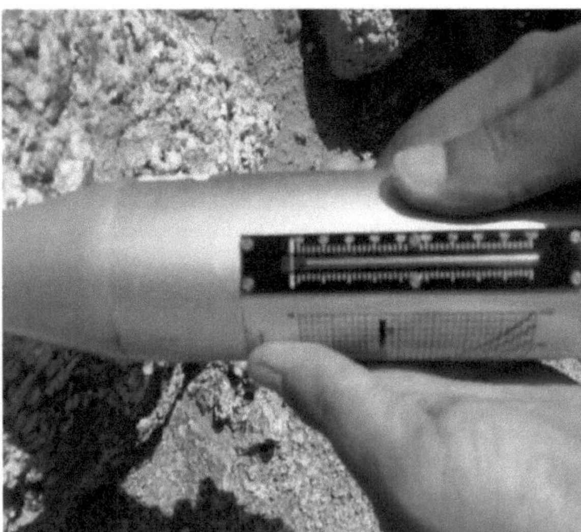

Figure 4.2 - Schmidt hammer, taken from Mohamad et al.(2011)

The Schmidt Hammer was developed in 1948 for non-destructive tests to determine the hardness of concrete, and was later used on rock materials (Katz et al., 2008).

al., 20[00)] . para a determining the properties of rock materials, the L and N type models can be used.

Type L has an impact energy of 0.735 N.m and is used in materials with less hardness or structural units with a thickness of less than 100 mm. It is the most commonly used in studies of rock masses and was used in this study.

The N-type Schmidt hammer has an impact energy of 2.207 N.m and is generally used to study concrete. The material to be tested must have a minimum thickness of 1000 mm and be well fixed (Fleury et al., 2012).

The uniaxial compressive strength of rock material can be determined using a portable sclerometer or Schmidt hammer (Figure 4.1). Studies by Hucka (1965), Deere & Miller (1966) and Dearman (1974) have indicated that R values, in combination with specific weight values, provide a good prediction of the uniaxial compressive strength of rock material.

CHAPTER 5

Characterisation of unstable situations

5.1 - Description of unstable situations

Catumbela's geographical layout is characterised by very uneven terrain, with steep slopes cut through by drainage networks and slopes with instability problems on which houses have been built in precarious situations in terms of stability.

There was an anarchic occupation of the territory by the inhabitants of the municipality of Catumbela, under the impotent gaze of the local state administration.

The increase in population and the disorganised construction of housing is directly related to the country's recent past - the civil war only ended in 2002.

The aim of this study is to inventory the situations of slope instability in this new municipality of Catumbela, which has only been in existence for three years, in order to study the behaviour of slope movements, specifically their dimensions, as well as their causes and consequences.

A study was carried out of 11 slopes located in the area of the city of Catumbela (Fig. 5.1), designated from Slope 1 (T.1) to Slope 11 (T.11) (Fig. 5.1), the vast majority of which correspond to excavation slopes.

All the slopes show active instability; vegetation is generally non-existent, with some slopes showing the presence of undergrowth, while two slopes showed moderate vegetation, characterised by the presence of small trees.

The main types of instabilisation identified were landslides, but there were also flows and landslides. The speed of movement was generally classified as very fast to fast.

Figure 5.1 - Image of the location of the slopes studied using Google earth (T. 1 - Slope 1; T.2 - Slope 2; T. 3 - Slope 3; T. 4 - Slope 4; T. 5 - Slope 5; T. 6 - Slope 6; T. 7 - Slope 7; T. 8 - Slope 8; T. 9 - Slope 9; T. 10 - Slope 10; T. 11 - Slope 11).

Almost all of the slopes studied were not stabilised. Small retaining walls can be seen on slopes 1 and 11. On slope 10, there are situations of instability related to the construction of houses that were built without observing the necessary technical requirements. The implementation of housing in the urban area of Catumbela often leads to changes in the geometry of the slopes or embankments, which can increase their inclination and/or increase the overloads at the top of the embankments, thus causing instability situations.

The main external causes of instability include: surface erosion, water infiltration, the application of overloads, an increase in the slope's inclination and the presence of animals that circulate at the top of the slopes. According to the study, the internal causes include: lithology, geological structure, increased water pressure and reduced soil resistance.

Slope 1

The geographical coordinates of Slope 1 are 12°25.31' South and 13°32.543' East. This slope is located in the Cambambi neighbourhood, along the road between Catumbela and Lobito, and has a height of around 14 m and a length of 125 metres. The face of the slope has a direction of N25°E and a slope of 80°NW - subvertical. The stratification of the layers present on slope 1 has the following geological coordinates: N25°E; 10° - 15° W.

On the slope there is a fault with a north-south orientation and a 20°E slope. It is 3 metres long, has an opening of around 10 cm and is essentially filled with clay. There are three main families of diacrasis, a sub-vertical family with a direction of N20°E, a sub-horizontal family parallel to the stratification and another with a direction of N70°W and a sub-vertical slope.

Slope 1 is made up of limestone, gypsum marl, sometimes with iron oxides, beach-sea terraces (at the top of the slope) and embankment material. The limestones are fine-grained and generally greyish to yellowish in colour. Marls are very fine grained and grey to brownish in colour. Terraces are located on the crest of the

39

slope, as well as cover and embankment material, varying in thickness from 0.5 to 2 metres; these materials may show signs of erosion. Schmidt hardness has average values of 26.8 and 10.8 for limestone and marl respectively.

The thickness of the limestone strata is generally between 0.14 and 0.50 metres, while the thickness of the marl layers varies between 0.05 and 0.60 metres (Fig. 5.2).

The most common types of instability are: falling blocks, some flows and more rarely planar landslides. The speeds of the instability movements that occur on the slope have been defined as very fast to fast. The rock blocks are roughly parallelepiped-shaped and have dimensions of 0.05 - 1.0 m x 0.08 - 0.6 m x 0.1-0.5 m.

As Slope 1 is adjacent to a motorway, instabilities have the effect of affecting users of the Catumbela - Lobito motorway (Calumba area), passers-by and also some homes (Fig. 5.2).

Figure 5.2- Slope 1 (Cambambi neighbourhood), where you can see the stratification of limestone and marl, the situations of falling blocks and the consequences of this instability.

Slope 2

Slope 2 has the geographical coordinates 12°25.325' South and 013°33.071' East and is located in the Poli neighbourhood. It is approximately 9.8 m high and 34 m long. The slope has a direction of N25°E and a slope of 75°-80°NW. The lithology of the slope comprises fine-grained limestone, grey, whitish to yellowish in colour, as well as fine-grained marl, yellowish to brownish in colour (Fig. 5.3). At the top of the slope there are houses built on top of roof deposits, the latter less than 0.5 metres thick.

The stratification has a slope between 15° and 25° W, the thickness of the limestone layers varies between 0.2 and 0.6 metres and that of the marls between 0.1 and 0.5 metres.

The vegetation on the slope was classified as non-existent to moderate. The instability situations observed were block falls and soil flows, the latter related to the cover deposits. The speed of the instability situations was considered very fast to moderate. The maximum size of the flow movements is approximately 20 metres[3]
. The average size of the rock blocks is 10 cm x 25 cm x 15 cm.

The instability movements affect users of the road linking the Poli neighbourhood to the Alto Niva neighbourhood, as well as the houses located near and on top of the slope. The external causes of the instability are related to the increase in weight on the slope, particularly at the top, as well as water infiltration and surface erosion. The main internal causes are the increase in water pressure and the decrease in the resistance of the terrain.

Figure 5.3 - Slope 2 (Bairro do Poli), where you can see the lithology present on the slope, the vegetation and some instabilised blocks. There is also the presence of a scar related to the detachment of blocks.

Slope 3

The geographical coordinates of Slope 3 are: latitude 12° 25.50' South and longitude 13°32.554' East. It is located in the Caputu neighbourhood, has a length of around 12 m and an approximate height of 7.4 m. Its direction is N50°W and its slope is 80°SW sub-vertical.

The dominant lithologies are fine-grained limestones with a yellowish to brownish hue and fine-grained marls with a dark yellow to brownish colour. There are gypsum veins filling in the discontinuities in the marls, varying in thickness from 5 mm to 2 cm.

The stratification has a slope of around 20°W, the thickness varies from 2 cm to 50 cm for the limestone layers and between 1 cm and 40 cm for the limestone layers.

marl. The most frequent types of instability are soil and debris flows and block falls (Fig.5.4). The average dimensions of the flow instabilities are approximately 40 cm wide, 100 cm long and 25 cm thick. The fallen blocks at the base of Slope 3 vary in size from 0.30 m x 0.50 m x 0.15 m (0.018 m^3) to 0.70 m x 0.90 m x 0.30 m (0.189 m).[3]

The movements of instability directly affect the houses in the immediate vicinity, particularly those built above the slope (Fig. 5.4). It should be noted that there has already been some damage to homes, particularly during periods of heavy rainfall, and there is currently no stabilisation work.

Figure 5.4 - Slope 3 (Caputo neighbourhood), where you can see the type of vegetation present, the flow movements, the percolation of water, as well as the size of the blockfalls.

Slope 4

Slope 4 has the geographical coordinates: 12°25.593' South and 13°32.662' East, and is located in the Caputo Baixo neighbourhood. It is 11.30 m high and 36 m long. It faces N60°E and has a sub-vertical slope.

The lithology on the slope is made up of limestone and marl with gypsum veins, sometimes oxidised. The limestones are fine-grained and predominantly greyish to yellowish in colour. The marls are very fine grained and greyish to brownish in colour. The layers of the marls have strands of gypsum, whitish in colour and 0.1 to 2 cm thick, filling discontinuities.

The thickness of the limestone strata is generally between 0.2 and 0.5 metres, while the thickness of the marls is mostly between 0.1 and 0.3 metres. The strata are orientated sub-horizontally.

At the top of the slope there is roofing material, with a maximum thickness of 50 cm, on which houses have been built.

The most common instability movements are block and debris falls and debris and soil flows. The predominant instability situation is block falls. The speed of the movements on the slope is generally considered to be very fast. The size of the unstable rock blocks is between 0.01-1.0 m x 0.01-0.6 m x 0.01-0.5 m. The elements related to debris falls have average dimensions of 10 mm x 7 mm x 10 mm.

The Schmidt hardness found for limestones was 30.84, while for marls it was 14.05.

The main consequences of the instabilisation were damage to homes and injuries. The internal causes of this instability are the lithology itself, the reduced resistance of the land, the geological structure - diaclasation and stratification, sometimes fault planes, and the external causes are the increased weight of the slope, surface erosion (as can be seen from the configuration of the slope in Fig. 5.5) and the presence of water.

Figure 5.5 - Slope 4 (Lower Caputo neighbourhood), where you can see the alternation of limestone and marl strata, as well as surface erosion.

Slope 5

Slope 5 is located in the Tata - Cal neighbourhood and has geographical coordinates of 12°25.582' South and 13°33.081' East. Slope 5 is about 30 metres long and 11 metres high, with slope deposits at its base about 5 metres high. Slope 5 has the geological coordinates: N40°W; 85°SW to sub-vertical. It is a slope with no vegetation, whose lithology is composed of limestone with yellowish tones, sometimes brownish, which indicates the presence of iron oxides, and reddish to dark brown colours (Fig. 5.6). The marls are generally whitish to grey. The thicknesses range from 2 cm to 70 cm for the limestones and from 1 cm to 30 cm for the marls. The stratification has a slope of approximately 15°W.

The most frequent types of instability are rock falls, and to a lesser extent debris falls and flows.

The average size of the flows is 10 cm wide, 18 cm long and 5 cm thick. The debris present in the flows has average dimensions of 14 mm x 12 mm x 8 mm. The unstable/unstabilised rock blocks reach dimensions of 1m long, 0.6m high and 0.5m thick.

The instability on the slope has been defined as active and could cause damage to homes, as well as causing human casualties when travelling or living in the vicinity of the slope. According to information from local residents, there have been several accidents related to falling blocks and/or debris onto children, which

43

have caused several injuries. Despite the various situations recorded, there are no stabilisation works and/or prevention measures in place.

The main causes of instability are related to lithological characteristics and fracturing, in particular diaclasation, the action of water and the frequent presence of animals, particularly goats and kids, which circulate on the top and face of the slope.

Figure 5.6 - Slope 5 (Bairro da Tata - Cal), visualisation of the stratification and presence of iron oxides.

Slope 6

Slope 6 is located in the Tata - Curva neighbourhood and has the geographical coordinates 12°25.658' South and 13°33.299' East. It is 30 m long and approximately 18 m high. The orientation of Slope 6 is N30°E; 80°SE sub-vertical.

As far as lithology is concerned, limestones and marls with iron oxides were found. The limestones present on the slope are generally brown to yellow and fine-grained, while the marls are grey to whitish and also fine-grained. Slope deposits and even debris can be seen at the base of the slope, sometimes up to 7.0 metres thick. The limestone strata on this slope have thicknesses of between 0.2 and 0.6 metres, while the marl intercalations have smaller thicknesses of between 0.1 and 0.3 metres. The stratification has an inclination of *20°W*.

It is actively unstable and its types range from falling blocks to debris flows (Fig. 5.7), with excavations at the foot of the slope posing a great risk to the houses built around it. At the bottom of the slope, the size of the instability reached over 10 metres[3]. The speed of movement was considered very fast to fast. The increase in the height of the slope, water infiltration and surface erosion correspond to the external causes of the instability, while the internal causes have to do with the lithology itself and the geological structure, as well as the increase in water pressure.

Figure 5.7 - Slope 6 (Bairro da Tata - Curva), where instability movements can be seen.

Slope 7

Slope 7 is located in the Vila Flor - Cametido neighbourhood and has the following geographical coordinates: 12°25.916' South and 13°33.006' East. It has the following geological coordinates: Direction N70°E; 80°S. The slope is 21.40 metres long and 10 metres high.

The lithology on the slope is made up of limestone and marl. The limestones are fine grained and predominantly greyish in colour. The marls are very fine grained and greyish in colour. The thickness of the limestone layers varies from 0.2 to 0.7 metres and the marl layers are between 0.2 and 0.6 metres thick. There are a number of fault planes, as can be seen in figure 5.8, the opening of which varies between 1 and 5 cm, their direction is approximately perpendicular to the stratification and they extend for around 1 to 8 metres.

The most frequent types of instability are falling blocks (debris and rocks) and soil flows (Figure 5.8). The speed of movement on the slope is considered to be fast. The rock blocks involved in instability situations vary in size from 1 cm x 2 cm x 7 cm to 1.0 m x 0.5 m x 0.4 m (Fig.5.8).

The consequences of the instabilisation are damage to houses and a Catholic chapel located near the foot of the slope. The internal causes of this instability are the lithology itself, the geological structure such as the diacrasis and fault planes, the decrease in the resistance of the land and the external causes are the increase in weight on the crest of the slope, the infiltration of water and surface erosion.

Figure 5.8 - Slope 7 (Vila Flor neighbourhood - Cametido), showing the consequences of instability that could affect residential buildings located close to the slope.

Slope 8

Slope 8 is located in the Chiúle neighbourhood and, with geographical coordinates of 12° 26.419' South and 13° 33.575' East, is 29 m long and 25 m high. Its direction is N80°E and its slope is 80°S to sub-vertical. The deposits at the base have a slope of 35°S.

The lithology present on the slope is made up of limestone, gypsum marl with calcite minerals. The limestones are fine-grained and greyish in colour. The marls are very fine grained and greyish to brownish in colour (Fig. 5.9).

The thickness of the limestones varies between 2 and 50 cm and the thinner marls vary between 5 cm and 30 cm. The geological structure is heavily eroded due to the presence of closely spaced diacrasis and strong surface erosion.

The most frequent types of instability are block falls, debris and soil flows and planar landslides. The speed of movement on the slope is considered to be very fast to fast. The size of instability situations related to fallen rock blocks and debris can reach more than 30 metres.[3]

The Schmidt hardness determined for limestones is 24.0. The causes of instability movements are the lithology, the decrease in resistance of the land, the geological structure - faults, and the external causes are water infiltration and vibrations. Instability situations can cause damage to homes and reach communication routes.

Figure 5.9 - Slope 8 (Chiúle neighbourhood), where anthropogenic action can be seen altering the geometry of the slopes.

Slope 9

Slope 9 is located in the São Pedro neighbourhood at 12° 26.911' South and 13° 32.447' East. It is 7-8 m high and 21 m long. Its direction is N30°W and its slope is 40° SW - sub-vertical.

The lithology present on the slope is made up of limestones - quite compact, fine-grained, marls with strands of gypsum, sometimes oxidised. The marls are very fine-grained and greyish to brownish in colour, with crystallisations of calcite and gypsum. The thickness of the limestone strata is generally between 0.2 and 0.5 metres, while the thickness of the marls is mostly between 0.1 and 0.3 metres. The stratification has an inclination of 20°W. The Schmidt hardness of the limestones is 28.6.

Covering material can be seen at the top and sides of the slope, with a maximum thickness of around 0.6 metres. There are also slope deposits at the base of the slope, which can reach a thickness of 1 metre.

It is a slope where the absence of vegetation stands out. The most frequent types of instability are block falls, debris and soil flows and planar landslides. No stabilisation work was found. The speed of movement on the slope was defined as fast to very fast. The maximum size of the instability situations related to fallen rock blocks is an average of 54 cm x 40 cm x 12 cm.

47

The consequences of instabilisation are damage to homes (Fig. 5.10), a school and a health centre. The internal causes of this instability are the lithology itself, a decrease in the resistance of the land, an increase in water pressure, the geological structure (stratification, faults), and the external causes are an increase in slope, an increase in weight, water infiltration, surface erosion and a decrease in the resistance of the land.

Figure 5.10 - Slope 9 (S. Pedro neighbourhood) where blocks have fallen down

Slope 10

The geographical coordinates of Slope 10 are: 12° 28.092' South and 13°31.882' East, it is located in the Luongo neighbourhood and has a direction of N40°W and a slope of 40°SW. It is about 120 metres long and approximately 15 metres high.

The lithology on the slope belongs to the upper Albian and shows a significant amount of ammonite fossils, as well as calcite crystallisation and gypsum minerals. The stratification on the slope is sub-horizontal.

It is a slope where vegetation is non-existent to rare. There are also slope deposits at the base of the slope, as well as cover deposits with a maximum thickness of around 1 metre. The most frequent types of instability are block falls, debris and soil flows and planar landslides. No stabilisation work has been carried out. The speed of movement on the slope was considered to be very fast to fast. The instability situations related to the rock blocks have average dimensions of 70 cm x 80 cm x 25 cm.

A Schmidt hardness value of 19.9 was determined for the limestones. The thickness of the limestone layers, which are white to greyish in colour, is generally between 0.2 and 0.6 m. The layers of marl are grey and range from 0.1 to 0.4 m thick.

The most serious consequences of instability on this slope could occur in the near future due to the construction taking place at the foot of the slope (Fig. 5.11), and instability could affect buildings that are currently under construction.

The internal causes of instability are the lithology itself, the geological structure and the action of water, and the external causes are the increase in the height of the slope, surface erosion, the application of overload at the top of the slope and the infiltration of water.

Figure 5.11 - Slope 10 (Luongo neighbourhood) This slope is evidence of human action in altering the initial geometry of the slopes.

Slope 11

The geographical coordinates of Slope 11 are: 12° 25.233' South and 13° 33.041' East, and it is located in the Alto Niva neighbourhood. The slope has a direction of N80°W and a slope of 70°S. The height of the slope is between 6.0 and 7.5 metres and its length is approximately 100 metres.

The lithology present on the slope is made up of limestones - quite compact, fine-grained and yellowish to brown in colour, marls with oxidised gypsum. The marls are very fine grained and greyish to brownish in colour. The strata have a slope of 20°W. The limestone strata vary in thickness from 3 to 60 cm and the marls from 1 to 45 cm.

The slope is generally free of vegetation, but cacti can be seen (Fig. 5.12) on its upper side.

The most common types of instability are falling blocks. There has been some stabilisation work, such as the construction of a wall on the eastern side of the slope. The speed of movement on the slope is considered to be very fast. The volume of all the instability situations related to fallen rock blocks in an instability event is approximately 5m3.

The consequences of the instabilisation are damage to homes and commercial buildings and traffic disruption on the road linking Bairro do Poli to Alto Niva. The internal causes of this instability are the lithology itself, the increase in water pressure, the geological structure (diaclasation and stratification), and the external causes are the excavations on the slope, the increase in loads at the top of the slope, vibrations and water infiltration (Fig.5.12).

Figure 5.12 - Presence of housing at the top of Slope 11 and falling blocks.

5.2 - Comparative study of instability situations

The study of instability situations revealed that the urban area of Catumbela is a region where there is the possibility of numerous slope/verge instability situations, since the characteristics detected on the slopes studied generally have a geological structure with several families of diacrasis that can be spaced apart. The relief in the area studied is somewhat rugged and most of the slopes are bare of vegetation, which is sometimes low to moderate, which promotes erosion processes. In addition to the natural features, anthropogenic action often chooses the slopes to open up roads, build housing and other engineering buildings, thus contributing to the occurrence of instability movements.

According to the results of using the worksheet, various comparisons can be made.

5.2.1 - Dimensions of the slopes studied

With regard to the size of the slopes, it was found that 3 slopes are longer than 100 metres, in this case slopes T 1, T10 and T11, which are 125 metres, 120 metres and 100 metres respectively. The remaining 8 slopes have lengths of less than 40 metres. As far as height is concerned, there is one slope with a value above 20 m (Slope 8), seven slopes with a height between 10 m and 20 m, and three with a height below 10 m. (Fig. 5.13).

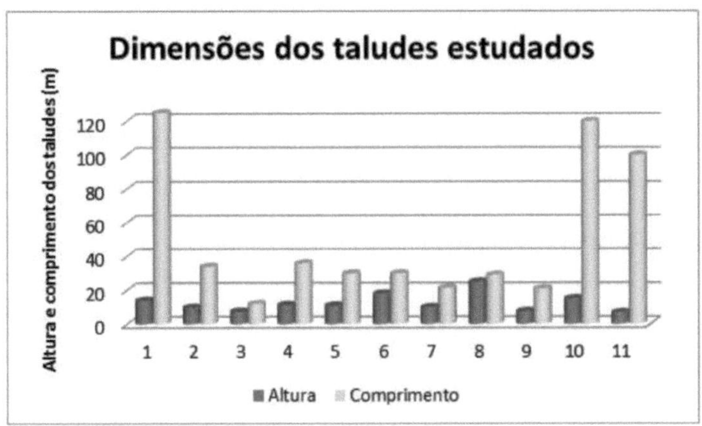

Fig. 5.13 - Slope dimensions.

5.2.2 - Lithology

The area studied is dominated by limestones and marls belonging to the Quissonde Formation, which originates in an outer platform environment and is made up of intercalations of limestones, marly limestones and marls with shell fragments at the base, lagemas and shell fragments in the middle and lagemas at the top. The formation is rich in macrofauna (ammonites, echinoids, gasteropods) (Buta-Neto, et al., 2006), which falls within the Albian (Alb.3), which falls within the Albian (Alb.3). The vast majority of the slopes studied show the occurrence of slope deposits and cover deposits. On Slope 1 it is possible to see Pleistocene marine terraces located at the top of the slope. Numerous ammonite fossils were found in the marly limestones on Slope 10.

5.2.3 - Vegetation

Of the eleven slopes studied, eight showed no vegetation cover, corresponding to 72.7 per cent, while three slopes (Slopes 2, 3 and 6) showed low to moderate vegetation (27.3 per cent), as can be seen in Fig. 5.14.

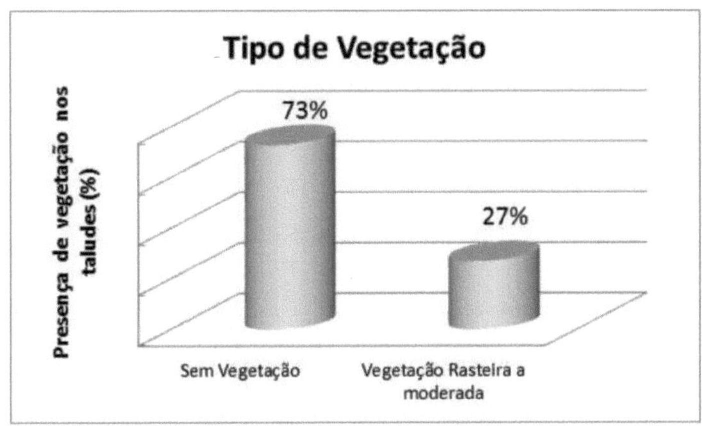

Fig. 5.14 - Distribution of vegetation on the slopes studied.

5.2.4 - Type of instability

The same types of instability occurred on practically all the slopes studied, with block falls standing out among them. Thus, in all eleven slopes (100%), rockfalls were observed. Soil and debris flows were observed on nine of the slopes (81.8%). Slopes 1, 8, 9 and 10, which correspond to 36.4% of the slopes studied, also showed instability due to planar landslides (Fig.5.15). Complex movements were observed on slope 5.

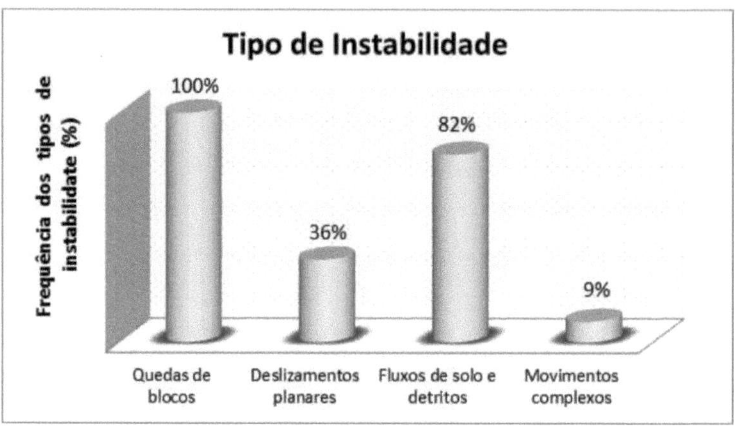

Figure 5.15) - Frequency of instability observed on slopes (%).

5.2.5 - State of activity, consequences and causes of instabilities

For all the slopes, the instability was considered to be active and was manifested by block falls, landslides, flows and also complex movements.

The instability situations observed have affected or could cause damage essentially to homes. Road embankments such as Embankments 1, 2 and 11 showed situations of instability that affected the roads, also affecting houses located at the top of the embankments, as can be seen in Fig. 5.2. On slopes 3 and 5, some victims were injured, particularly children (Fig. 5.16).

Instability movements have various internal and external causes, so for the former, lithology and geological structures have a decisive influence on instability occurrences at the eleven sites studied. In eight (72.7 per cent) and seven (63.6 per cent) slopes, the increase in water pressure and the decrease in soil resistance were also considered to be internal causes (Fig. 5.17), and it should be noted that in all slopes the predominant lithology consists of interspersed limestone and marl.

Water infiltration is an external cause of instability present in all the slopes, for seven of which the increase in slope overloads and surface erosion were defined as external causes of instability, while in another four slopes the increase in slope inclination was considered; in another three the external causes were vibrations, the increase in slope height and the presence/activity of domestic animals (Fig. 5.18).

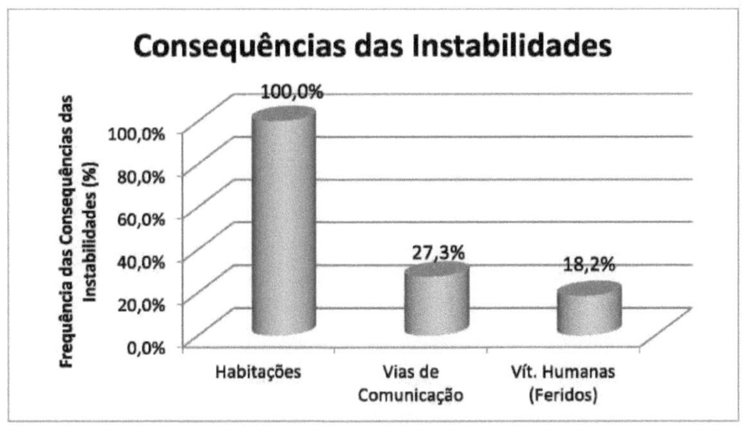

Figura 5.16   - Consequences of the instability situations observed on the slopes studied.

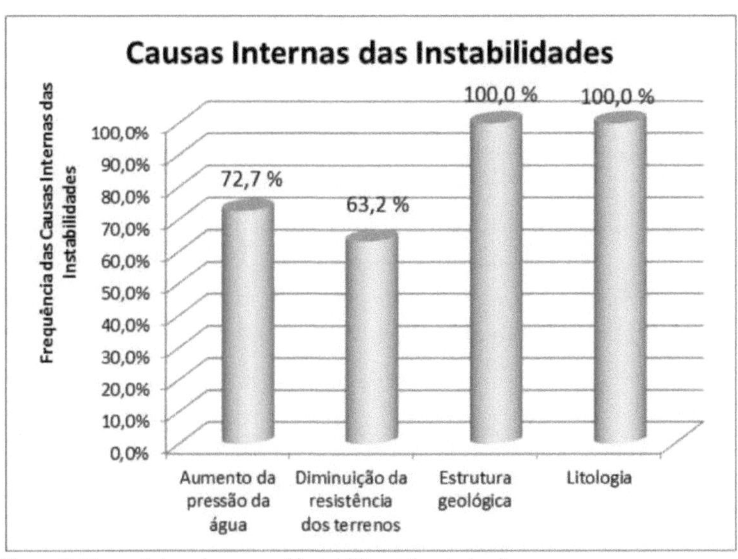

Figura 5.17   - Internal causes of instability at the study sites.

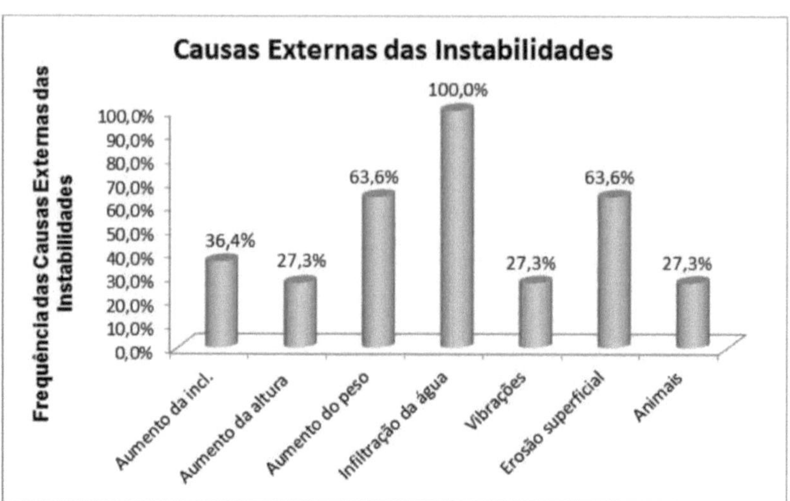

Figura 5.18   - External causes of the instability detected on the slopes.

5.3 . Application of the Rockfall Hazard Rating System

Problems associated with falling blocks of rock can arise on motorways. These unstable movements can affect road users and vehicles, causing damage and casualties.

The Rockfall Hazard Rating System (RHRS) was applied to three rock excavation slopes, Slopes 1, 2 and 11. The various parameters that make up the RHRS classification were used, for each of which weighted values were defined according to the results of the work carried out in situ. The weighted point classification ranges from 3 to 81 points (Table 4.2). It should be noted that in order to assign scores to the parameters of the RHRS method, the values of the imperial system of the original Pierson et al. classification (1991) were converted to the units of the International System (metric system).

5.3.1. - Slope height

The height of the slopes studied was defined, considering that each one consists of a single section.

The height of Slope 1 is around 14 metres, including the thickness of the Pleistocene marine terraces and the cover deposits that are located on top of it. According to the height value, a weighted value of 9 points was assigned. For slopes 2 and 11, which have heights of 9.8 m and 6 to 7.5 m respectively, a weighted value of 3 points was set.

5.3.2. - Ditch retention capacity

Due to the lack of retention ditches on the various slopes, it was considered that there is no capacity to retain the various instability situations, so the weight value for this parameter is equal to 81 points.

5.3.3. - Medium risk for vehicles

The three slopes have different locations. Slope 1 is located on the so-called "Catumbela/Lobito highway", which connects Cambambi - Catumbela and the Calumba neighbourhood - Lobito, passing through the Santa Cruz neighbourhood - Lobito, in an area of heavy traffic. On Slope 1, rock blocks are likely to fall, especially during periods of heavy rainfall, which can be seen in the considerable number of blocks on the carriageway or on its edges. In dry weather, falling blocks are not frequent, but the possibility of vehicles and people travelling along the Catumbela/Lobito highway being hit is not ruled out.

According to the measurements taken *on site, there is traffic on* the carriageway, in in average terms, 80 vehicles per hour.

The speed limit was 30 km/h. Slope 1 is 125 metres long, so the average risk for vehicles (AVR) on the Catumbela - Lobito motorway is 33.3%. In light of the above, a weight value of 8 points was assigned.

Slopes 2 and 11 are located in areas with less road traffic, with an average of 20 vehicles per hour. Slope 2 is 34 metres long and the speed limit is 30km/h. The AVR value is 2.3%, giving a weight value of 3 points.

For T11, with a length of 100 metres and a speed limit of 25 km/h, the AVR was 8%, so the weight value was set at 3 points.

5.3.4. Decision visibility distance

Drivers circulate on the roads in the vicinity of Slopes 1, 2 and 11 at speeds generally in excess of 30km/h, and in some situations it may not be possible to avoid the unstable materials coming from the slopes, since the width of the roads is reduced, with the roads located on Slopes 2 and 11 being only 6 and 5 metres wide respectively, and there is also heavy traffic, particularly on Slope 1.

The decision visibility distance for the three slopes was defined using equation 4.4 and the results for Slope 1, Slope 2 and Slope 11 were 84.6 per cent, 27.7 per cent and 30.8 per cent respectively, so the weight value for Slope 1 is 9 points, while for the other slopes the weight values are 81 points.

5.3.5. Track platform width

On Slopes 1, 2 and 11 the width of the road platform is 12.53 m, 6 m and 5 m respectively, and it should be noted that the verges are not paved. The weighted classification was 5 points for Slope 1, while for Slopes 2 and 11 the weighted value was 81 points.

5.3.6. Geological characterisation

In view of the geological survey carried out, Case 2 was considered, which is related to the structural condition and differentiation in erosion rate of the material making up the slopes.

Slope 1 has a very high structural erosion condition, mainly due to the presence of the marl layers, which is why the classification of 54 points was considered; for Slopes 2 and 11, in terms of structural condition, the presence of high surface erosion was observed, which made it possible to define the weight value of 27 points for each of the slopes. As for the differentiation of the erosion rate, it was found that the materials making up the slopes show high to extreme differences, so a classification of 54 points was set for all the slopes considered.

5.3.7. Block size

The unstable blocks on Slope 1, which could reach the road, have a size that allows them to be considered with a weight value of 27 points.

The average volume of fallen rock material on Slopes 2 and 11 is smaller than that on Slope 1, which made it possible to define a weight of 3 points for Slope 2 and 9 for Slope 11, given that in the latter there was an instability event where the volume of fallen blocks was around 5 m^3 .

5.3.8. Climatic conditions and presence of water on the slopes

The rainfall conditions were considered to be moderate, although the annual rainfall values are approximately 270 mm (Costa, 2013), the rainfall events are of great intensity and therefore cause most of the instability situations that occur on the slopes considered. Bearing in mind the rainfall characteristics, the weight values were set at 9 points for the different slopes.

5.3.9. Historical record of the fall of the block

For the different slopes studied, the frequency of block falls was defined as occasional. Although there are no records of the volume of instability situations in recent years, rockfalls are common during periods of heavy rainfall. The weight value assigned to the parameter of the historical record of falling blocks was assumed to be 9 points.

5.3.10. Rockfall Hazard Rating System values

In order to summarise the scoring carried out using the RHRS classification, the results for slopes 1, 2 and 11 are shown in Table 5.1.

Table 5.1 - Scoring results for the parameters considered in the RHRS classification.

Parameters			Score		
			Slope 1	Slope 2	Slope 11
Slope height			9	3	3
Ditch efficiency			81	81	81
Medium risk for vehicles			8	3	3
Decision visibility distance			9	81	81
P width	track reform		5	81	81
Nature Geological	Case 1	Structural condition			
		Rock friction			
	Case 2	Structural Condition	54	27	27
		Differentiation in erosion rate	54	54	54
Block size			27	3	9
Number of block falls per event					
Presence of water on the slope			9	9	9
History of falling blocks			9	9	9
GRAND TOTAL			265	351	357

The RHRS classification does not indicate which stabilisation methods should be used, taking into account the sum of the weighted values of its parameters. According to Pierson et al. (1990) and Hoek (2007), slopes with RHRS classification values below 300 are not a priority in terms of stabilisation, while slopes with RHRS classification values above 500 should be subject to urgent stabilisation actions.The results showed that Slopes 2 and 11, with RHRS classification values of 351 and 357 respectively, require stabilisation interventions in the short term, while for Slope 1, whose RHRS classification value is 263, these same interventions can be carried out in the medium term.

5.4 - Recommendations

From the study carried out on the instability present on the slopes and taking into account their location, it is necessary to recommend some measures that should be taken as a matter of urgency;

1. Carry out a detailed geotechnical study of the slopes that are most susceptible to instability movements and where these can cause damage and casualties;

2. Sensitise people to the risks to which they are exposed;

3. In view of what has been said in this paper and considering the results of the RHRS classification assessment, and also taking into account the causes of slope instability in the study area, gabion walls should be built in the most fractured and altered areas of the slopes and unstable cover deposits, or protective nets should be installed, especially on the slopes located along the roads (Slope 1, Slope 2 and Slope 11), in order to protect road users and vehicles, as well as passers-by. For the infrequent situations of large, unstable blocks of rock, the use of nailing can be used.

4. Another possible measure is to reduce the slope of rocky slopes and cover deposits, which could be removed, thus avoiding situations of flow movements and falling material.

5. With the approval of the restriction letters, coercive prohibition measures are needed to prevent people from settling in certain areas, as well as the creation of a strong body of inspectors who will constantly appeal to people to comply with the national legislation in force.

6. Publicise the municipal master plan so that everyone is aware of it and knows about it, so that it can be fully complied with and defended.

CHAPTER 6

CONCLUSIONS

The work carried out in this dissertation included inventorying and characterising the instability situations that occur in the Catumbela municipality, particularly in sensitive areas with high population density, such as the neighbourhoods of Cambambi, Poli, Caputu, Tata, Vila Flor, Chiúle, São Pedro and Luongo. Many of the buildings in these neighbourhoods are located on slopes, thus obstructing water lines, and are also located at the top of slopes, causing an increase in load.

The study area has a climate defined as arid or semi-desert. The relief is in the form of escarpments and hills that are generally separated by depressions, sometimes with high slopes. The dominant soils on the slopes are generally thin and can sometimes be clayey.

The eleven slopes studied (T1 to T11) are made up of carbonate sedimentary rocks from the Lower to Middle Cretaceous of the Benguela Basin (Figs. 3.8 and 3.9), which are part of unit Alb3, and whose lithofacies are similar to those of the Quissonde Formation. The lithological materials found on the slopes are essentially limestone, marl limestone and marl. Terrace deposits belonging to the Pleistocene are sometimes present and are visible on the right bank of the Catumbela River, near the Cambambi neighbourhood, Alto Niva and in the Morro do Galo neighbourhoods.

This study carried out a survey of instability situations in the neighbourhoods of Cambambi, Poli, Caputu, Tata, Vila Flor, Chiúle, São Pedro, Luongo and Alto Niva, defining a worksheet in order to study existing instability situations. Various parameters were considered, such as the geometric characteristics of the slopes, the lithology, morphology, geological structure and geological formations present on the slopes. The hardness of the lithological materials present on some of the slopes was defined. The results obtained were also analysed and interpreted, taking into account the parameters on the worksheets.

The Rockfall Hazard Rating System (RHRS) was used to determine the potential risk situations associated with instability on three road embankments and to prioritise prevention and mitigation actions.

The vast majority of slopes (72.7 per cent) have no vegetation cover, with only three slopes (27.3 per cent) having it.

Instability was defined as active, with rockfalls occurring on all the slopes studied. Debris and soil flows were also recorded on around 82 per cent of the slopes, while planar landslides occurred on 4 of them.

The main consequences of instability are damage to homes located close to the slopes, as well as disruption to communication routes. In the last three years, four people have been injured as a result of instability.

Schmidt hardness values ranged from 19.9 to 30.84 for limestones and from 10.8 to 14.05 for marls. It should be mentioned that in some marl strata it was not possible to carry out Schmidt hardness determinations, as values below 10 were recorded.

The most significant internal causes of instability are the lithology (intercalation of limestone and marl layers) and the geological structure (diaclasation, stratification and sometimes fault planes). External causes include the action of water, increased overloading caused by constructions at the top of slopes and surface erosion.

The results of applying the RHRS classification revealed total values of 263, 351 and 357 for Slopes 1, 2 and 11 respectively, which made it possible to define interventions with higher priority characteristics for Slopes 2 and 11, with less urgent stabilisation and prevention work for Slope 1.

The research made it possible to carry out an integrated analysis of the information obtained, thus being able to identify the main similarities and differences in the instabilities present on the slopes studied.

The work has made it possible to acquire knowledge that can be an integral part of a more in-depth analysis of the instability situations that occur on the slopes and slopes of the municipality of Catumbela, making a valuable contribution to the study of environmental processes and also to land use planning.

BIBLIOGRAPHICAL REFERENCES

AHI (2014). Affordable Housing Institute. http://affordablehousinqinstitute.org/bloqs/us/. Accessed on 15 January 2015.

AMC (2012). Municipality Profile - 2012 - 2013. Catumbela Municipal Administration.

AMC (2014). 4th quarter report - 2013. Catumbela.

Andrade, P.G.S. (2008). Classification of movements in natural and excavation slopes. Coimbra: Department of Earth Sciences - FCTUC, 39p.

Andrade, P.G.S (2008). Study of the main characteristics of discontinuities. Coimbra: Department of Earth Sciences - FCTUC, 43p.

Andrade, P.G.S. (2009). Mass Movements: slides from the course on Analysing and Managing Natural Risks. Coimbra: DCT FCT University of Coimbra.

Andrade, P.G.S. (2012). Mass Movements: Support slides for classes in Natural Risk Analysis and Management. Coimbra: Department of Earth Sciences, University of Coimbra.

Araújo, A.G.; Guimarães, F. (1992).Geologia de Angola: Notícia Explanativa da carta geológica à escala 1: 1.000.000. Geological Survey of Angola. Publisher. Luanda: Geological Survey of Angola. 135p.

Araújo, A.G.; Prevalov, O.V.; Guimarães, F. R.; Kondratiev, A. L; Tselikov, A. F.; Khodiev, V. L. (1998). Mineral resources map on a scale of 1:1000 000. Republic of Angola: IGEO.

Ayala-Carcedo, F.J.; Andreu, F.J.P. (2006). Manual de Ingeniería de Taludes. Madrid: Geological and Mining Institute of Spain. 456p.

Buddeta, P.; Panico, M. (2002.). Il metodo "Rockfall Hazard Rating System" modificato per la valutazione del rischio da caduta massi. Technical & Environmental Geology. Journal of Technical & Environmental Geology, 2, pp. 3-13.

Bastos, A. (1912). Monograph of Catumbela. Lisbon: Lisbon Geographical Society. 100p.

Bastos, M.J. (1999). Structural stability in the safety of open-cast quarries - earth massifs. VISA - Technical Communications, pp. 1-4.

Bloom, A.L. (1970). Surface of the Earth. São Paulo, Editora Edgard Blúcher Ltda., 184p.

Braathen, A.; Blikra, L. H.; Berg, S. S.; Karlsen, F. (2004.). Rock-slope failures in Norway; type, geometry, deformation mechanisms and stability. Norwegian Journal of Geology, 84, pp. 6788.

Buta-Neto, A.; Tavares, T. D.; Quesne, D.; Guiraud, M.; Meister, C.; David, B.(2006). Synthèse préliminaire des travaux menés sur le bassin de benguela (Sud Angola): implications sédimentologiques, paléontologiques et structurales. Africa Geoscience Review , 13, pp. 239 - 250.

Carreto, A.P. (1989). Slope stabilisation techniques. Ingenium - Revista da Ordem dos Engenheiros, June, pp. 63 - 73.

Carvalho, G.S. (1957). Some Problems of the Quaternary Terraces of the Coast of Angola. INQUA (International Association for the Study of the Quaternary), Barcelona, 11 p.

Carvalho, G.S. (1961). Some problems of the Quaternary terraces of the Angolan coast. Boletim de Minas, Angola, 2, pp. 5-15.

Carvalho, G.S. (1963). Problemas da sedimentologia das praias do litoral de Angola. Garcia de Orta : journal of the Junta das Missões Geográficas e de Investigações do Ultramar. Lisbon. Vol. 11,2, pp. 291-313.

Carvalho, P.A. (1991). Motorway embankments: Guidance for diagnosing and solving problems. São Paulo: IPT, 338p.

Coates, D.R. (1977). Landslide. Reviews in Engineering Geology. The Geological Society of America, vol. III, 278 p.

Coelho, A.G. (1979). Cartographic analysis of slopes for urban planning. GEOTECNIA - Revista da Sociedade Portuguesa de Geotecnia, 26, pp. 75 - 89.

Cruden, D.M. (1991). A simple definition of a landslide. Bulletin of the International Association of Engineering Geology, 43, pp. 27-29.

Cruden, D.M. (2011). The Working Classification of Landslides: material matters. 2011 Pan-An CGS - Geotechnical Conference. Canada: University of Alberta, Edmonton, Alberta, Canada, pp. 1 - 7.

Cruden, D.M.; Varnes, D.J. (1996). Landslides. Investigation and Mitigation. Special Report 247. Ed. Keith Tuner and Robert Schuster. Transportation Research Board. National Research Council. National Academy Press, Washington D.C., 247p.

Cruden, D.; VanDine, D. F. (2013). Classification, Description, Causes and indirect effects - Canadian Technical Guidelines and Best Practices related to Landslides: a national initiative for loss reduction. Geological Survey of Canada, Open File 7359, 22 p.

Cruz, J.R. (1940). The Climate of Angola. Elemento de climatologia. Lisbon, 96p.

Dearman, W.R. (1974). Weathering classification in the characterisation of rock for engineering purposes in British practice. Bulletin of international Association of Engineering Geologists. Vol. 9, pp. 33-42.

Deere, D.U.; R. P. Miller (1966). Engineering classification and index properties of rock. Tech. Report Air Force Weapons Lab, New Mexico, pp. 65-116.

Department of Roads and Highways of the State of São Paulo (D.E.R.E.S.P) (1991). Manual de Geotecnia: Taludes de Rodovia; Orientações para Diagnósticos e Soluções de seus problemas. São Paulo: Technological Research Institute: IPT Publications 1843, 206p.

D.R. (2013). National Strategic Plan for Water 2013 - 2017. *Diário da república, I série, n.°22 - Decreto Presidencial n.°9/13*, pp. 250-284.

Dikau, R.; Brunsden, D.; Schrott, L.; Ibsen, M. (1996) - Landslide Recognition. John Wiley & Sons, Chichester, 251

p.

Diniz, A.C. (2006). Mesological Characteristics of Angola: Description and Correlation of Physiographic Aspects, Soils and Vegetation of Angolan Agricultural Zones. Lisbon: IPAD. 56p.

Diniz, A. C. (1998). Angola, the Physical Environment and Agricultural Potential, 2ª Revised Edition, 59p.

Dyminski, A.S. (2010). Class notes for the course: Slope Stability. In A. S. Dyminski, Noções de Estabilidade de Taludes e Contenções, UFPR, Brazil, 28 p.

Favaretti, M. (2011). Environmental Geotechnics - Landslides Classifcation. Padova: University of Padova.

Feio, M. (1960). The raised beaches of the Lobito and Baía Farta region. In G. de Orta, Angola Pleistocene, Vol. 8, pp. 357 - 370.

Feio, M. (1981). The relief of south-west Angola. Geomorphological study. *Memórias da Junta* de Investigação Científica do Ultramar, 32p.

Fernandes, N.F.; Guimarães, R. F.; Gomes, R. A.; Vieira, B. C.; Montegomery, D. R.; Greenberg, H. (2001). Geomorphological Constraints of Slope Landslides: Evaluation of Methodologies and Application of a Model for Predicting Susceptible Areas. Brazilian Journal of Geomorphology , 2, pp. 51 - 71.

Flageollet, J.C.; Weber, D. (1996). Fali. In R. Dikau, D. Brunsden, L. Schrott and M. L. Ibsen (eds.), Landslide recognition - Identification, movement and courses. Chichester: John Wiley & Sons Ltd., pp. 13-28.

Fleury, S.V.; Olkowski, G. F.; Kurokawa, E. (2012). Proposed methodology for differentiating hard and soft rock in audits using the Silver Schmidt Sclerometer. National Technical Meeting on Public Works Auditing - ENAOP - Palmas/TO, pp. 1-12.

Galvão, C.F.; Silva, Z. (1972). Geological Map, sheet no. 227-228 Lobito, scale 1: 100 000. Luanda.

Gerscovich, D. M. (2009). Slope stability. Rio de Janeiro: Faculty of Engineering (FEUERJ) - Department of Roads and Foundations.

Giresse, P.; Hoang, C. T.; Kouyoumontzakis, G. (1987). Analysis of vertical movements deduced from a Geochronological Study of Pleistocene deposits Southern Coast of angola. (P. i. Britain, Ed.) Journal of African Earth Sciences , 2, pp. 177 -187.

Goodman, R.E.; Bray, J.W. (1976) - Toppling of rock slopes. Speciality Conference on Rock Engineering for Foundation and Slopes, Boulder, Colorado. ASCE, vol. 2, pp. 201 - 234.

Benguela Provincial Government (GOB) (undated). FIMA - Angola International Mining Fair. Benguela has Minerals . Benguela, Benguela, Angola.

Guerra, A.J.; Jorge, M. D. (2012). Everyday Geomorphology. Soil degradation. GeoNorte , 4 (Special), pp.116 -135.

Guerreiro, H.J. (2000). Underground Exploration of Marble - Geotechnical Aspects; Course Conclusion Work (Master's Degree in Geo-Resources - Geotechnics Area). Technical University of Lisbon, Portugal, 133p.

Guiraud, M.; Buta-Neto, A.; Quesne, D. (2010). Segmentation and differential post-riftbuplift at the Angola margin as recorded by the transform-rifted Benguelaq and obliqúe to-orthogonal-rifted Kwanza basins. Marine and Petroleum Geology, vol.27, pp. 1040-1068.

Hansen, A. (1984). Landslide Hazard Analysis. D. Brunsen and D. B. Prior (eds.). Slope Instability. New York: Johan Wiley & Sons. pp. 523-602.

Hart, D. J.; Wang, H. F. (1995). Laboratory measurements of a complete set of poroelastic moduli for berea sandstone and Indiana limestone. J. Geophys. Res. 100, pp. 17741 - 17751.

Highland, L. M.; Bobrowsky, P. (2008). The Landslide Handbook - A Guide to Understanding Landslides. (P. R.

Rogério; J. J. Aumond, Trads.) Brazil: Mary Kidd, 156p.

Hoek, E. (2007) - Practical rock engineering. http://www.rocscience.com/hoek. Accessed on 15 January 2015.

Honrado, J.; Martins, F.; Calejo, M. J.; Dos Santos, H. K.; David, J. M. (2010). Angola's National Irrigation Master Plan: A synthesis of studies. The Engineering of Hydro-agricultural Reservoirs: current situation and future challenges - APRH Technical Conference, Luanda, 17 p.

Hucka, V. (1965). A Rapid Method of Determining the Strength of Rocks in Situ. Int. J. Rock Mech. Mining Sei., 2, pp. 127-134.

Hungr, O.; Evans, S. G.; Bovis, M. J.; Hutchinson, J. N. (2001). A Review of the Classification of Landslides of the Flow Type. Enviromental & Engineering Geoscience, VII, pp. 221 - 238.

Hutchinson, J.N. (1988) - General report: morphological and geotechnical parameters of landslides in relation to geology and hydrogeology. Proc. 6th International Symposium on Landslides. Ed. C. Bonnard. Balkema, Rotterdam, vol. 1, pp. 3 - 26.

Infanti Jr., N.; Filho, N. F. (1998) Processes of Surface Dynamics. In: Engineering Geology. Santos, A, M dos; Oliveira, S. N A. de B. São Paulo: Brazilian Association of Engineering Geology, pp. 131-152.

Júnior, A.P.; Longo, O. C. (2010). Analysis of mass movements in urban areas: the case of the Dom Giocondo neighbourhood. Energy, Innovation, Technology and Complexity for Sustainable Management. Niterói, RJ, Brazil: V Congress of Excellence in Management, pp. 1-19.

Kamussel, C. (2009). kamussel.forums-free.com. provincia-de-benguela-geografia-e-historia de http://www.benguela.net/benguela/catumbela.asp:http://kamussel.forums-free.com/provincia- de-benguela-geografia-e-historia-t197. 2431 p. Accessed on 05 September 2014.

Katz, O.; Reches, Z.; Roegiers, J.-C. (2000). Evaluation of mechanical rock properties using a Schmidt Hammer. International Journal of Rock Mechanics and Mining Sciences, 37, pp. 723 - 728.

Lamas, P.C. (1998). The slopes of the south bank of the Tagus River - Geomorphological evolution and failure mechanism. Lisbon: Doctoral dissertation in Geotechnics, specialising in Engineering Geology, FCT/UNL, 379p.

Lawrence, A.; Pierson, C. E.; Robert Van Vickie, R. P. (1993). Rockfall Hazard Rating System. Washington DC: Federal Highway Administration. 113p.

Li, Z.; Huang, H.; Xue, Y.; Yin, J. (2009). Risk assessment of rockfall hazard on highways. Georisk: Assessment and Management of Risk for Engineered System and Geohazards, 3, pp. 147 - 154.

Lopes, E.S.S. (2006). Dynamic spatial modelling in Geographic Information Systems - an application to the study of mass movements in a region of the Serra do Mar in São Paulo. PhD Thesis in Geosciences and Environment. Institute of Geosciences and Exact Sciences of UNESP (IGCE/UNESP), Rio Claro, 314p.

Luciano, P. (2008). Quantitative Risk Assessment of Rockfall Hazard in the Amalfi Coastal Road. Barcelona: Universitat Politécnica de Catalunya, 90p.

Marangon, M. (2006). Unit 03 - Geotechnical retaining walls, part 1. In M. Marangon, Topics in Geotechnics and Earthworks, pp. 109 -123.

Martins, A.A. (2007). Contribution of geomorphology to the preparation of municipal master plans (PDMs): application in the municipality of Barcelos. Accessed on 15 May 2014, from http://repositorium.sdum.uminho.pt/handle/1822/7262: http://hdl.handle.net/1822/7262

Martins, J.B. (2002). Foundations. University of Minho, Braga.

Martins, R.J.; Moreira, P.N.; Carlos, S.P.; Neto, E.P.; Alzira, M.P.; Teixeira, J. (2009). The Linear Sampling Technique Applied to Compartmentalisation Studies of Rock Massifs in Northern Portugal. Portugal, 10p.

Mattos, K.C. (2009). Instabilisation processes on road slopes in sandy residual soils: Study on the Castello Branco highway (SP. 208), Km 305 to 313 (Master's thesis). São Carlos, 111 p.

Mendes, P.E. (2012). Time Series Analysis of Component 2 of the Water Audit Project on the Angolan Side of the Cubango/Okavango River Basin. Luanda, 61 p.

Meneses, B.M. (2010). Landslides, a natural catastrophe. Report presented to the Faculty of Social and Human Sciences of the New University of Lisbon, 16p.

MINUA (2006). Report on the general state of the environment in Angola. Environmental Investment Programme, Ministry of Urbanism and Environment. Luanda, 325p.

Mohamad, E.T.; Gofar, N.; Amin, M. M.; Isa, M. F. (2011). Field Assessment of Rock Strength by Impact Load Test. Malaysian Journal of Civil Engineering, 23 (1), pp. 105-110.

Mouta, F.; O' Donnell, H. (1933). Carte Géologique de l' Angola, Esc. 1/2.000.00. Exolicative Notice - Ministry of the Colonies . Lisbon.

Nogueira, N.A. (2010). Stability analysis of artificial slopes. Lisbon, 144p.

Novotny, J. (2013). Varnes Landslide Classification (1978). Addis Ababa: Addis Ababa University, Ethiopia, 21 p.

Oliveira, R. (1979). Natural and excavation slopes. Lisbon: New University of Lisbon.

Ortis, J.M. (undated). Terrain Movement. In F. J. Ayala-Carcedo, Manual de Ingeniería Geologica. Spain: ITGE, 626p.

Pereira, M. (2001). Stereographic Projection in Structural Geology - Problems and Solutions. Angola Minas, 12 (IITrim.), pp. 22-25.

Pierson, L.A., Davis, S. A.; Van Vickle, R. (1990). Rockfall Hazard Rating System - Implementation Manual, Federal Highway Administration (FHWA), Report FHWA-OR-EG-90-01, FHWA, U.S. Dep. of Transp, 11 p.

Pinho, A.C.; Carvalho, F. F. (2010). Oil prospecting, exploration and production in Angola. The role played by oil companies. In. J.M. Cotelo Neiva, A. Ribeiro, L. Mendes Victor, F. Noronha e M. Magalhães Ramalho (Eds.) *Ciências Geológicas - Ensino e Investigação e sua história*. Portuguese Association of Geologists, Lisbon, pp. 61-70.

Pires, R.; Gardete, D.; Luzia, R. (2012) - Application of the "Rockfal Hazard Rating System" to the slopes of the EN353 motorway in Idanha-a-Nova. 13° Congresso Nacional de Geotecnia, Lisbon, 17-20 April. Portuguese Geotechnical Society, pp. 1-16.

Ramos, T.M. (2009). Geomechanical tests on metasedimentary rocks from the Mina das Aveleiras massif (Mosteiro de Tibães): comparison, potential and limitations (Master's thesis). Porto, Portugal: ISEP - Instituto Superiorn de Engenharia do Porto, 219p.

Rapp, A. (1961). Recent development of mountain slopes in Karkevagge and surroundings, northern Scandinavia: Geografiska Annaler, v. 42, pp. 71-200.

Rodrigues, M.; Zêzere, J.L. (2003). The Fonte Nova landslides (Alcobaça). Geomorphological characterisation and associated risks. III Seminar on geological resources, environment and land use planning, pp. 1 - 8.

Romariz, C. (1970). Carbonate sedimentary rocks - General considerations and economic importance. Naturália, 131 p.

Roquinaldo, F. (2013). Biography as social history - the Ferreira Gomes clan and the worlds of enslavement in the

South Atlantic. Varia História, 29, pp. 679 - 695.

Russel, C.; Santi, P.; Higgins, J. (2008). Modification and Statistical Analysis of the Colorado Rockfall Hazard Rating System. Association of Environmental and Engineering Geologists 50th Annual Meeting Programme with Abstracts, Los Angeles, CA, p. 119.

Santos Preira C.M.; Chaminé, H. L.; Vieira, A. R.; Teixeira, J.; Gomes, A.; Fonseca, P. E. (2005). Structural geology and geotechnics of the granite massif of Alto da Cabeça Santa (NW Portugal): implications for the management of the geo-resource of the for the management of the geo-resource of the. (E. d. castro, Ed.) Revista de Xeoloxia galega e do Hercínio Peninsular, 30, pp. 39 - 56.

Santos, G.; Zacarias, I. (2010). Research into land disputes and conflicts and ways of resolving them. Luanda, 136p.

Saraiva, A.L.; Andrade, P. G. S. (2003). Temporal evolution of natural and excavation slopes. Engineering Geology and Geological Resources. Coimbra University Press (Eds) Martim Portugal Ferreira (Coord), vol 1, pp. 355-365.

Sassa, K. (1985). The mechanism of debris flows. In: Proceedings of XI International Conference on Soil Mechanics and Foundation Engineering, San Francisco, 3, pp. 1173- 1176.

Serralheiro, R. P.; Monteiro, F. G.; Sousa, P. L. (undated). Irrigation in Angola from the perspective of rural development, pp. 55-73.

Sharpe, C.F. (1938). Landslides and related phenomena. New York: Columbia University, 171p.

Sieira, A.C. (2009). Geosynthetics and Tyres: an alternative for slope stabilisation. ENGEVISTA ,11, pp. 50-59.

Silva, J.L. (2011). Occupation of hillsides and landslides in Córrego do Euclides, Recife - PE. Monograph presented to the Department of Geographical Sciences, UFPE. July.

Small, R.J.; Clark, M. J. (1982). Slope and Weathering. Cambridge: Cambridge University Press, 112p.

Tavares, T. (2005). Ammonites et échinides de l'Albien du basin de Benguela (Angola). Systématique, biostratigraphie, paléogeéographie et paléoenvironnement. Unpublished thesis. Dijon, France: Université de Bourgogne.

Tavares, T.; Meister, C.; Duarte-Morais; M. L.; David, B. (2007). Albian ammonites of the Benguela Basin (Angola): a biostratigraphic framework. South African Journal of Geology, vol. 110, pp. 137-156.

Teixeira, A.D. (2012). Sustainability assessment of surface landslides. Use of physically based mathematical models in the Tibo Basin, Arco de Valdevez (Master's thesis). University of Porto, 118p.

Teixeira, M. (2005). Slope movements - occurrence factors and inventory methodology. Portuguese Association of Geologists, Geonovas, 19, pp. 95-106.

Tominanga, L.K. (2009). Slides. In L. K. Tominanga, J. Santoro;R. Amaral, Natural Disasters: Knowing to Prevent. 1.ª Edition ed., São Paulo,197p.

Torquato, J.R. (1978). Geotectonic outline of Angola. II centenário da Academia de Ciências de Lisboa, Lisbon Sep. Estudos de Geologia e Paleontologia e de Micologia, pp.123-149.

Vallejo, L. I.; Ferrer, M.; Ortuno, L.; Oteo, C. (2002). Geological Engineering. Madrid, Spain: Prentice Hall.

Varnes, D.J. (1958) - Landslides Types and Processes. Special Report29: Landslides and Engineering Practice. Ed. E.B. Eckel. HRB, National Research Council, Washington D.C., pp. 20-47.

Varnes, D.J. (1978) - Slope movement types and processes. Special Report 176 - Landslides: Analysis and Control. Ed. R.L. Schuster, R.J. Krizek. TRB, National Research Council, Washington D.C., pp. 11-33.

Varnes, D.J. (1984) - Landslide hazard zonation: a review of principies and practice. UNESCO, Paris, 63p.

Vedovello, R.; Macedo, E. S. (2007). Slope Landslides. In R. F. Santos, Environmental Vulnerability: Natural Disasters or Induced Phenomena? Brasília, 192p.

Victorino, A.D. (2012). Geokwanza: Development of a WebSIG for the Sedimentary Geology of the Kwanza Sedimentary Basin (Project Work). Lisbon, 55p.

W.P./W.L.I - UNESCO (1993). Multilingual Landslide Glossary. International Geotechnical Society, Canadian Geotechnical Society. Richmond: Bi Tech Publishers Ltd.

Zaruba, Q.; Mencl, V. (1976). Engineering Geology. Elsevier, Amsterdam, 504p.

Zêzere, J.L. (1997). Slope Movements and Geomorphological Hazards in the Region North of Lisbon. Lisbon: Doctoral dissertation in Physical Geography presented to the Faculty of Letters of the University of Lisbon, 575p.

Zêzere, J.L. (2000). The Classification of Slope Movements: Typology. Activity and Morphology. Lisbon: Centre for Geographical Studies, 35p.

Zêzere, J.L. (2005). Slope dynamics and geomorphological risks. Lisbon: Centre for Geographical Studies, 129p.

Printed by Books on Demand GmbH, Norderstedt / Germany